Particle Mechanics

Other titles in this series

Linear Algebra
R B J T Allenby

Mathematical Modelling
J Berry and K Houston

Discrete Mathematics
A Chetwynd and P Diggle

Vectors in 2 or 3 Dimensions
A E Hirst

Numbers, Sequences and Series
K E Hirst

Groups
C R Jordan and D A Jordan

Probability
J McColl

In preparation

Ordinary Differential Equations
W Cox

Analysis
E Kopp

Statistics
A Mayer and A M Sykes

Calculus and ODEs
D Pearson

Modular Mathematics Series

Particle Mechanics

C D Collinson

School of Mathematics
University of Hull

T Roper

School of Education
University of Leeds

Elsevier Ltd.
Linacre House, Jordan Hill, Oxford OX2 8DP
200 Wheeler Road, Burlington, MA 01803

Transferred to digital printing 2004

British Library Cataloguing in Publication Data
A catalogue record for this book is available from the British Library

ISBN 0 340 61046 8

1 2 3 4 5 95 96 97 98 99

Typeset in 10/12 Times by
Paston Press Ltd, Loddon, Norfolk

Contents

Series Preface

This series is designed particularly, but not exclusively, for students reading degree programmes based on semester-long modules. Each text will cover the essential core of an area of mathematics and lay the foundation for further study in that area. Some texts may include more material than can be comfortably covered in a single module, the intention there being that the topics to be studied can be selected to meet the needs of the student. Historical contexts, real life situations, and linkages with other areas of mathematics and more advanced topics are included. Traditional worked examples and exercises are augmented by more open-ended exercises and tutorial problems suitable for group work or self-study. Where appropriate, the use of computer packages is encouraged. The first level texts assume only the A-level core curriculum.

Professor Chris D. Collinson
Dr Johnston Anderson
Mr Peter Holmes

Preface

This book has been written not only for readers who have already studied mechanics at pre-University level but also for those who are studying mechanics for the first time. Accordingly the subject is introduced *ab initio* with a full discussion of its foundations, including our intuitive ideas about motion. Emphasis is placed on understanding and consequently readers already familiar with the subject will find new insights into some of the earlier material, particularly the more subtle aspects which are inappropriate for discussion in a pre-University course. Understanding cannot be harnessed to the solving of problems without the development of manipulative skills. These two complementary aspects of learning are covered in the Tutorial Problems throughout the text and in the exercises at the end of each Section and Chapter. The book is suitable for mathematicians and physicists alike and for others wishing to gain further insight into the subject. SI units are used and, whenever possible, the symbols used are those recommended by the Royal Society.

1 • The Role of Mathematical Modelling

The main features of mathematical modelling are discussed and related to the scientific method of Descartes. Mechanics is defined as the study of mathematical models of motion, divided into kinematics and dynamics, and particles are introduced as mathematical models of bodies moving in circumstances in which it is appropriate to neglect their size. A survey of our intuitive ideas about mechanics is intended to encourage readers with no or little practical laboratory experience. A distinction is made between impulsive and continuous forces and emphasis placed on the different effects they have on the motion of a particle. The linear approximation of a function is introduced and used to derive mathematical models of the tension in a stretched string and the drag force on a body moving through a resisting medium. The discussion throughout this chapter is naive, attempting to build on the reader's experience and observations. No formal definitions are given. Nevertheless, the foundations are laid for a more formal discussion later.

1.1 General Discussion

What is mathematics? Perhaps one way to answer this question, or to side step it – depending on your point of view – is to say that mathematics is what mathematicians do. If you were to investigate the interests of a hundred mathematicians employed in industry, in the public services and in academic institutions, you would quickly become aware of a very wide spread of interests. Some, for example, would be interested in mathematical structures for their own sake, striving to generalize existing structures, forging links between seemingly disparate structures and deepening understanding whenever possible; they would be called pure mathematicians. Others would be interested in analysing data, designing experiments, devising methods for the quality control of industrial processes; they would be called statisticians. Still others would be interested in using mathematical structures to describe and enhance their understanding of the world about us; they would be called applied mathematicians. In fact the distinction between these three areas of interest is far from clear cut and to use them to categorise mathematicians is often less than helpful. The applied mathematician searching for mathematical structures with which to describe some specific phenomena often motivates the discovery of new mathematical structures or the reinterpretation of existing ones. The pure mathematician who uses algebraic structures to describe and investigate geometrical results, for example by the use of cartesian coordinates, is engaged in much the same activity as the applied mathematician.

The use of mathematical structures to describe and so enhance understanding of specific aspects of the world about us is called mathematical modelling. The mathematical structures used are said to constitute a mathematical model of the phenomena being described. The choice of mathematical model will depend on our experience of the phenomena, experience which can be gained from the results of experiment, the analysis of observations, conjectures and previously established mathematical models of similar phenomena. Mathematical modelling plays a central role in many disciplines, for example in physics, chemistry, the biological sciences, economics, geography and the social sciences. The difference between the arts and the sciences is that science is concerned with just those phenomena which can be modelled mathematically. The chemist who performs an experiment to find a reaction between substances, models the observed reaction as a symbolic equation. The biologist models birth and death processes as differential equations. The manager models the activities of different sections of a business by means of weighting factors which are then used in the formulae constructed to determine the allocation of resources to those sections. The engineer models the relationship between the tensions and thrusts in the struts of a bridge as a system of algebraic equations.

Having formulated a mathematical model of some observed phenomena it is possible to manipulate the mathematics and to interpret the results of the manipulation in order to predict the occurrence of some new phenomena. It is this element of prediction which excites the interest of the academic and which is of practical use to society. If the predictions are found to occur in reality then the mathematical model is validated, that is it is found to be satisfactory. However once a prediction fails to agree with reality the mathematical model is shown to be incomplete and must be replaced by a more refined model. No mathematical model can ever be accepted as being perfect. When a mathematical model has been validated by testing its predictions against reality, those predictions themselves extend the original observations on which the mathematical model was based and thus might suggest possible modifications to the model. As experimental and observational techniques improve, the accuracy of the information obtained increases and this again might suggest refinements to an existing mathematical model. The important point to bear in mind from this discussion is that the process is never ending; existing mathematical models will be superseded, eventually, by more sophisticated ones. Nevertheless it is also important to understand that it is pointless to use sophisticated mathematical models when more naive ones will give predictions and results compatible with the accuracy of any experiments and observations made.

The whole process of setting up a mathematical model of observed phenomena and then testing the predictions against reality is often referred to as the "scientific method" and is usually attributed to the French natural philosopher René Descartes (1596–1650), after whom cartesian coordinates are named.

As an example of mathematical modelling consider two cul-de-sacs which are used regularly for car parking when events are staged at a nearby stadium. Before you read on, try the following problem.

TUTORIAL PROBLEM 1.1

(Part 1)
What measurements might you have to make in order to estimate the number of cars parked in each cul-de-sac at time *t*; how could you use these estimates to predict which cul-de-sac fills with cars first?

In considering this problem you are actually establishing a mathematical model for the number of cars parked and then making a prediction based on the model. Here is a possible approach, the cul-de-sacs being denoted by P_1 and P_2. The number of cars parked in each of P_1 and P_2 at time t will depend on the rates at which cars enter the two cul-de-sacs; these rates can be modelled as two real numbers c_1 cars/second and c_2 cars/second. The actual values of c_1 and c_2 can only be found by

Fig 1.1 Two cul-de-sacs.

measurement. If at time $t = 0$ there are no cars parked in either cul-de-sac then we can predict that the number of cars parked in P_1 and in P_2 at time t seconds will be $c_1 t$ and $c_2 t$ respectively, it is being assumed that the rates c_1 and c_2 are constant, so that $c_1 t$ and $c_2 t$ should be thought of as estimates. If $c_1 > c_2$ then the number of cars parked in P_1 will, at all times, be greater than the number of cars parked in P_2. We would predict therefore that the first cul-de-sac will fill up before the second and this prediction might well be of use to the policeman controlling traffic to the event. Unfortunately this prediction may not occur in reality because no account has been taken of the fact that once a cul-de-sac fills up then the number of cars in that cul-de-sac remains constant, i.e. the formula ct for the number of parked cars does not remain valid. The mathematical model needs modifying by taking into account the total number of cars which can be parked in P_1 and in P_2. Let these numbers be n_1 and n_2 respectively. The first cul-de-sac will fill up when $c_1 t = n_1$, i.e. at time $t = n_1/c_1$. Similarly the second cul-de-sac will fill up at time $t = n_2/c_2$. The required prediction can now be made by determining which is the smaller of the two ratios n_1/c_1 or n_2/c_2.

In the process of modelling it is often necessary to make approximations. Usually the phenomena being modelled are very complex and so, in order to achieve a reasonably simple mathematical model, it is necessary to neglect certain aspects of the phenomena and to model only those dominant aspects which, perhaps intuitively, are of importance. For example, in the parking model under discussion the number of cars which can be parked in each cul-de-sac is very difficult to estimate. These numbers depend on the length and curvature of the cul-de-sac, the

length and breadth of each individual car, the spaces left between parked cars, etc. In order to achieve a simple mathematical model suppose that the cul-de-sacs are of lengths l_1 and l_2 and are straight, and that the cars are each of length l. This length l could be taken to be the average length of all the cars sold in some period plus a factor to account for the space left between parked cars. Then the total number of cars which can be parked in P_1 and P_2 is $n_1 = l_1/l$ and $n_2 = l_2/l$. The prediction as to which cul-de-sac fills up first was found to depend on which is the smaller of the two ratios n_1/c_1 or n_2/c_2. This prediction can now be reinterpreted as depending on which is the smaller of the two ratios l_1/c_1 or l_2/c_2. Notice that there is no need to find the actual value of l so long as it is assumed that the average lengths of cars parked in each cul-de-sac are equal.

TUTORIAL PROBLEM 1.2

(Part 2)
An expensive restaurant is situated in the first cul-de-sac P_1. How might this information change the model in the text? How might it change your model?

Perhaps in considering this problem you have concluded that the restaurant will attract some wealthier supporters for a pre-event meal and that the average length of the cars parked in P_1 will therefore be greater than that of the cars parked in P_2. This difference must be taken into account. A difficulty arises here because one can never know before a particular event that the restaurant will be used and so influence the parking of cars. At best one can estimate the probability that the restaurant will be used and then the mathematical model becomes a probabilistic model rather than a deterministic model.

TUTORIAL PROBLEM 1.3

(Part 3)
How did your model compare to the one suggested in the text? Were the same variables considered? What kinds of assumptions were made in both cases? Look again at the model in the text. Are there any assumptions that are not explicitly stated, that are hidden? Are there any such assumptions in your model?

Occasionally the whole process of mathematical modelling becomes inverted so that the understanding of the mathematical structures being used is obtained from the phenomena being modelled. One example of such an inversion is based on the fact that in the usual mathematical model of an electrical circuit the current flowing in the circuit satisfies a certain differential equation. If the circuit is very complex then the differential equation is very complex and so difficult to solve. However it is a trivial matter to measure the actual current flowing in the circuit at different times and these measured values, when plotted against time, yield the graph of the solution to the differential equation. This method of "solving" differential equations is the basis of the electronic analogue computer. It is possible to feed the current directly into an oscilloscope and hence to display the graph of the solution directly.

Phenomena from very different disciplines are often modelled by the same mathematical structures. The phenomena are then analogues of each other. As an example of this situation consider the economist's problem of constructing the cheapest road system between given towns. The given towns can be modelled as points on a plane and the roads can be modelled as straight line segments lying on the plane.

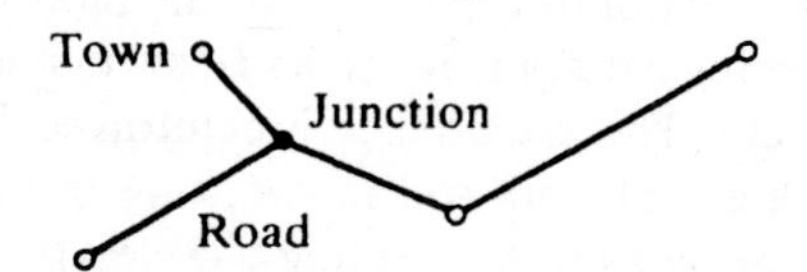

Fig 1.2 The road system joining towns.

If the cost of constructing each kilometre of roadway is the same at all points, and if the cost of constructing junctions is neglected, then the cheapest road system will be that of minimal total length. The mathematical determination of such a road system is very difficult indeed. Now consider two parallel plates of glass a distance d apart. Suppose a map of the region containing the towns is etched onto one plate of glass and suppose that perpendicular rods are placed between the plates of glass at the locations of the towns. If this is dipped into a soapy solution then bubbles will form between the rods, these bubbles being perpendicular to the plates of glass. Viewed from the plate of etched glass these bubbles will form a network of line segments connecting the towns. The effect of surface tension is to minimize the surface area of the bubbles, so that the bubbles will move until their area, which is just the constant d times the total length of the network, is a minimum. Hence in its final configuration the network will be a network of minimal total length and will therefore correspond to the cheapest road system between the towns. This is an example of a physical analogue of a problem in economics.

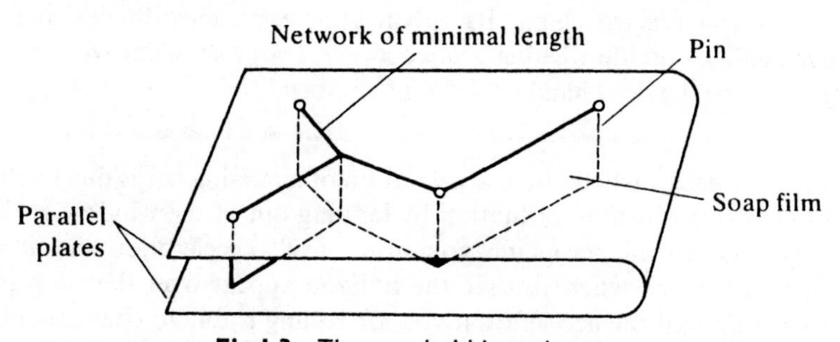

Fig 1.3 The soap bubble analogue.

1.2 Intuitive Ideas about Motion

Mechanics is concerned with motion and in particular with the study of mathematical models of motion. Those mathematical models concerned with general motions are formulated as the laws of mechanics. Mathematical models describing particular motions of particular systems are then constructed within the framework of these laws. Although the laws of mechanics have evolved from the painstaking analysis of observations and experiments, they can be understood in terms of our intuitive ideas about motion, provided that we reflect on those intuitive ideas very carefully and augment them with sensible conjectures.

Childhood play gives us a wealth of experience of the motion of specific bodies – pulling wheeled toys, pushing carts, throwing balls, swinging, sliding, flying kites, twanging elastic bands, etc. This experience is reinforced later in almost every aspect of our everyday life – playing sports, working in the kitchen or garage, driving, sitting on trains, or buses or cars, riding bicycles, taking lifts, to give but a few examples. The naive ideas of moving faster and slower, apparent for example in running games, is formalized later by the concept of **speed**, the quantity measured by the speedometers seen in cars and other vehicles. The importance of the **direction of motion** is learnt through the twisting and turning movements, made in order to escape capture by a friend.

When travelling in a car along a straight road you will have noticed that you can "feel" when the speed of the car is increasing – you detect a reaction between the seat and your back. Similarly, even if the car is travelling with a constant speed, you can "feel" when the car is turning a corner. If you are sitting with your shoulder in contact with the appropriate side of the car you detect a reaction between the side of the car and your shoulder. The sensations are entirely similar and in both cases the car is said to be **accelerating** so that, intuitively, acceleration occurs whenever either the speed or the direction of motion is changing.

TUTORIAL PROBLEM 1.4

The next time you are travelling as a passenger close your eyes and devise methods for detecting an increasing speed, a decreasing speed, a right hand bend and a left hand bend, each bend being taken at a constant speed. Suppose your left shoulder is in contact with the side of the vehicle in which you are travelling. Based on your experience in deriving the above methods decide whether the reaction on your shoulder will increase when turning the right hand or the left hand bend.

You can easily decide whether the acceleration of a moving car is due to changing speed or changing direction of motion by looking out of the window at the road and hedgerows. In an aeroplane you also "feel" acceleration but it can be something of a shock when you see the horizon appear over the wing tips and suddenly realize that the acceleration you are feeling is due to changing direction rather than changing speed!

The fact that acceleration occurs in two seemingly very different circumstances suggests that when describing the motion of a body its speed and direction of motion should be modelled together as a single vector quantity. This is done by introducing the concept of **velocity** so that the previous two statements of the conditions for acceleration to occur can be combined into one, namely that acceleration occurs whenever the velocity is changing in magnitude or direction or, indeed, both.

To talk of the speed of a body, as has been done above, is in fact ambiguous because if a body is rotating then different points of the body will have different speeds! This difficulty disappears for a given body if, when modelling the motion of the body, it is appropriate to neglect its size. Such a body is said to be modelled as a **particle** and will be referred to simply as a particle. This book is concerned solely with the motion of particles.

The concept of a particle is very important and is perhaps best appreciated by an example. Consider the motion of a tennis ball whilst in flight. The radius of the ball is very small compared to the total distance travelled by the ball and so the path of the tennis ball could be drawn as a simple curve with the actual size of the ball neglected. When discussing the motion of the tennis ball along this path it would seem reasonable to model the ball as a particle. Predictions based on this model would be open to validation. When the tennis ball hits the ground its direction of motion will change by an amount which is determined by the spin which the player imparts to the ball when it is struck by the racquet. This change in direction takes place whilst the ball is in contact with the ground and therefore the distance travelled by the ball during this phase of the motion is not much different to the radius of the ball itself. In such circumstances there could be no justification for modelling the tennis ball as a particle when discussing this phase of the motion. The radius and, indeed, other physical properties of the ball would have to be included in the model.

Velocity and acceleration are properties of a moving particle and are of importance in that branch of particle mechanics called **kinematics**, that is in the study of mathematical models describing the observed motion of a given particle. The simplest possible motion of a particle is motion in a straight line with constant speed. Such motion is called **uniform motion**. Since neither the speed nor direction of motion of the particle is changing, the particle is not accelerating. Thus uniform motion of a particle can also be specified as motion with zero acceleration or with constant velocity. Notice that uniform motion includes a state of rest, as a special case when the speed is zero.

Before proceeding further with this discussion of our intuitive ideas about motion it is important to dispel a commonly held misconception about uniform motion, namely, that in order to maintain the motion of a particle moving on a straight line with constant speed a force has to be applied to the particle in the direction of the straight line. You must, at some time, have pushed an object along the floor and, having released it, observed it to move in a straight line but with decreasing speed. The object eventually comes to rest and so it is certainly true that in order to maintain a constant speed you would have to continue to push the object along the floor. Little surprise therefore that the above misconception is commonly held.

You have possibly also noticed, however, that the smoother and more polished the floor the less the force with which you have to push the object in order to maintain its constant speed. This suggests that the force with which you push the object is not directly maintaining the constant speed of the object but is rather counteracting some other force acting between the floor and the object, a force which when acting alone causes the object to lose speed. This force is called the force of friction between the floor and the object; its magnitude depends on the nature of the surface of the floor and object. The smoother and more polished the floor, the less the force of friction. Now we have to make a sensible conjecture; if the floor were to be completely smooth the force of friction would vanish and then the object being pushed along the floor would, when released, indeed move in a straight line with constant speed. Generalizing, motion in a straight line with constant speed – uniform motion – is the natural motion of a particle, and requires no nett force to maintain it. Any force which you have experienced as being necessary to maintain a uniform motion is, in fact, counteracting some other force in order to obtain a zero nett force. To test these assertions experimentally you must have access to a completely smooth surface. The surface of an ice rink is very nearly smooth but technology can provide even smoother surfaces, for example by eliminating actual contact with a cushion of air between the surface and the object moving on it.

Accepting that if the nett force acting on a particle is zero then the motion is uniform, do we have any intuitive understanding of what forces can act on a body and how these forces are related to the non uniform motions of a particle? The study of mathematical models, describing the relationship between the forces which are observed to disturb the uniform motion of a particle and the actual motions they cause is a second branch of particle mechanics, called **dynamics**.

You may remember having to force(!) a stubborn friend to move out of your way by giving him or her a push. Such a push is an example of a force. Pulling the string attached to a toy cart results in a force which moves the cart. Gravity results in the downward force so often associated with a falling apple and so, in order to overcome gravity, a flying kite must also experience some sort of upward force or "lift". Stretching an elastic band between your hands results in forces which will move your hands together unless opposed in some way. Our experience of these and other forces indicates that a force has both a magnitude and a direction and so will be modelled as a vector. For example pushing harder will cause your friend not only to move out of your way but to actually fall down – pushing certainly has a magnitude associated with it. Similarly pushing has a direction associated with it – witness the different directions in which you can make your friend move out of your way!

When teeing off in golf, the golf ball is initially at rest and it is the force exerted by the head of the golf club as it strikes the golf ball which causes the ball to move off the tee. This force is exerted whilst the club head is in actual contact with the ball, it therefore acts for only a very short period of time; such forces are described by the adjective **impulsive**. Another example of an impulsive force is the force exerted on a moving ball when it is kicked or struck by a hockey stick. The effect of this is to change both the speed and direction of motion of the ball. Since speed and direction of motion are modelled together as velocity it would seem that an impulsive force, modelled as a vector, is related to a change in velocity. Our experience might well suggest that doubling such an impulsive force will double the change in velocity, so that if a particle is initially at rest then doubling the magnitude of the impulsive force acting on it will double the speed with which it starts to move. This is indeed the case; the change in the velocity of the particle is proportional to the impulsive force and, conversely, the impulsive force is proportional to the change in the velocity. This is not the whole story. Consider the impulsive forces with which you would have to kick a tennis ball and a cannon ball in order that each should start to move from rest with the same speed. Our intuitive understanding of this situation warns us to investigate it by performing a thought experiment rather than risk injury by actually kicking the cannon ball! The impulsive force to be applied to the tennis ball will, of course, be considerably less than that to be applied to the cannon ball. We deduce that each body has a property which determines the impulsive force required to produce a given change in the velocity of the body. This property is modelled as a real number called **mass**, chosen so that the required impulsive force is proportional to it. With this choice the mass of the tennis ball will be considerably less than the mass of the cannon ball. Combining the two different aspects of the discussion given in this paragraph we see intuitively that

- the impulsive force which acts on a particle is proportional to the product of the mass of the particle and the resulting change in the velocity of the particle.

Many forces are not impulsive but act continuously, for example the force which moves a toy cart, the gravitational force and the frictional force between the floor and an object moving on it. When such a force is acting on a particle it will disturb the natural uniform motion of the particle. Research shows that there is another common misconception, namely that the force is in the direction of the resulting motion of the particle and that doubling the magnitude of the force will double the speed. In other words the misconception is that a continuously acting force has the same effect on a particle as has an impulsive force. This view was certainly held before the time of Newton and is quite false. It was one of the reasons for the lack of progress made before Newton in the understanding of motion. Since an impulsive force results in a finite change in velocity then a continuously acting force must surely result in a continuous change in velocity and therefore an acceleration. This was the breakthrough made by Newton, so that

- the force which acts continuously on a particle is proportional to the product of the mass of the particle and the resulting acceleration of the particle.

It is clearly important to distinguish between continuous and impulsive forces. It is conventional to omit the adjective continuous when discussing the former but to include the adjective impulsive when discussing the latter. Hence in what follows the word **force** alone will always refer to a continuously acting force.

In this section no attempt has been made to give any formal definitions – they will all appear later. No doubt readers who have already studied mechanics at pre-University level will have found the discussion naive, but it has emphasized the importance of thinking about and being aware of the commonplace motions and forces which are all about us. More formal experience, for example in a physics laboratory, is useful but is not essential to the study of the material to follow. Some readers having no formal background in the subject might find it useful to spend some time investigating mechanics with the Leeds Mechanics Kit, or with one of the other kits now available. Videos and computer simulations are also available, see the Appendix.

EXERCISE ON 1.2

1. Devise an experiment whereby the mass of a particle is modelled as a real number.

1.3 What is Particle Mechanics?

Particle mechanics is the study of mathematical models describing the observed motion of a given particle (kinematics) and those describing the relationship between the forces which are observed to disturb the uniform motion of a particle and the observed motions they cause (dynamics). Here the word particle is used to describe any physical body whose size can be neglected when modelling the particular motion of the body under consideration.

Suppose that the motion of a given particle has been observed for some specified period of time. Kinematics, being concerned solely with modelling the observed motion during this period of time, cannot predict the subsequent motion of the particle because there can be no way of knowing whether the state of motion may change after the initial period of observation. In contrast, dynamics does predict the subsequent motion, provided that the forces acting on the particle are known. For this reason the role played in particle mechanics by kinematics is quite minor compared to that played by dynamics. Three reasons for the importance of particle mechanics will become apparent as the subject develops. These are summarized below:

- particle mechanics has wide applicability to observations on the earth and experiments in the laboratory (terrestrial mechanics) and to the understanding of planetary motion and other motions within the solar system (celestial mechanics)
- for any body of finite size, or system of bodies, there exists a point whose motion is that of a particle; this point is called the centre of mass
- the mathematical models of the motion of bodies of finite size are derived from particle mechanics by techniques of integration.

TUTORIAL PROBLEM 1.5

Take a tennis racquet or some other non uniform object which is convenient to handle and throw. Try to predict what will happen if the tennis racquet is thrown by one member of the tutorial group to another so that it rotates. Perform this experiment and describe the observed motion. Find the balance point of the racquet and mark this point clearly, for example by sticking a disc of coloured paper on the racquet. Repeat the experiment and observe the motion of the disc and the motion of the racquet relative to the disc. What do you notice? Can you explain what you have observed?

TUTORIAL PROBLEM 1.6

Wrap a narrow strip of paper around a pencil at its mid point and secure with sellotape. Place the pencil on a horizontal table parallel to one edge and flick the pencil, perpendicular to its length, at its midpoint. Do this several times and describe the motion of the pencil and of its midpoint in particular. Repeat this experiment but now flick the pencil at one end, again in a direction perpendicular to the length. Describe the motion of the midpoint and the motion of the pencil relative to this point. Can you explain what you have observed?

As stated in the last section, mathematical models concerned with general motions are formulated as the laws of mechanics. The laws of mechanics considered here are those due to Sir Isaac Newton (1642–1727). The study of mathematical models of motion constructed within the framework of these laws is called **Newtonian mechanics** or **classical mechanics**. In the final chapter of this book an example is given of a prediction based on classical mechanics which is not validated by observation. An alternative formulation of the laws of mechanics is that due to Albert Einstein (1879–1955) and the study of mathematical models of motion constructed within the framework of these laws is called **relativistic mechanics**.

Care has to be taken in describing the relationship between classical and relativistic mechanics. To say that relativistic mechanics is a refinement of classical mechanics is to deny the revolutionary nature of Einstein's mathematical model of space and time; nevertheless classical mechanics is obtained as an approximation to relativistic mechanics for particles moving with speeds much less than the speed of light. To say that classical mechanics has been superseded by relativistic mechanics is to deny the continuing importance and general applicability of classical mechanics to all motions, excepting those involving speeds approaching the speed of light.

Both of the above theories have led in the past to the prediction of new and amazing phenomena. For example, application of classical mechanics to the irregularities in the orbit of the planet Uranus enabled the French theoretical astronomer Leverrier to predict, in 1846, the presence of a hitherto unknown planet. This planet was observed by Galle a year later, lying within one degree of the location predicted by Leverrier; you may know it as the planet Neptune. Application of relativistic mechanics enabled Einstein to predict, in 1905, the existence of nuclear energy almost fifty years before its use to generate electricity.

The continued success of classical mechanics is apparent in the technology which surrounds us, from the design of bridges to the design of supersonic aircraft, from the understanding of the absorption of oxygen into a plant's root system to the understanding of ice accretion on overhead transmission lines. The continued success of relativistic mechanics must be linked to that of two other major theories of the twentieth century, general relativity and quantum mechanics. Again many predictions could be quoted, for example the prediction of the existence of neutron stars by Oppenheimer and Sneider in 1939 more than twenty years before their role as "black holes" was discussed and the prediction of the existence of many elementary particles, for example quarks and the Higg's boson, long before particle accelerators could be constructed powerful enough to yield evidence for their existence.

The success of classical and relativistic mechanics must rank as one of the greatest achievements of mankind.

1.4 Modelling Space and Time

The concept of time surely has its origin in naturally occurring events. Observation of the rhythm of the seasons will have led to an estimate of the length of the year as a number of days, the day being the obvious interval between two sunrises. The advent of agriculture necessitated a more accurate "calibration" of the seasons and this was achieved, for example, by observing the motion of the sun. The length of the year was then specified by observing midsummer's day – the day on which the sun attains its greatest height in the sky. The phases of the moon will have led to the month and the rising and setting of the sun to the division of the day into two intervals, day and night. Notice the ambiguity in the use of the word day. Further domestication of the human race necessitated the further subdivision of the day, and this was done by observing the position of the sun or, equivalently, of a shadow, perhaps using a primitive sundial.

Instruments were developed for measuring the interval of time between the occurrence of two events, for example by counting the number of swings of a pendulum, or the number of oscillations of a caesium atom, between the occurrence of the two events. Thus intervals of time are modelled mathematically by real numbers and these real numbers are measured by instruments known as clocks. The real number which models a given interval of time will depend upon the standard unit of time used to calibrate the clock. Here the unit of time used will be the second. The time at which some event occurs is then simply the interval of time between the occurrence of some prescribed event, which is used to define the "origin" of time, and the occurrence of the event itself. For example the time of day is the interval of time between midnight and the present.

Different clocks, calibrated in seconds, often give different readings. This can be for one of two reasons. The first is that one of the clocks is faulty in which case it must be repaired or replaced. The second is that the clocks are not synchronized, i.e. they are using different prescribed events to define the origin of time. The important assumption made in the classical model of time is that once two identical clocks are synchronized then they will remain synchronized even when the clocks are in relative motion. Such a model is said to assume the existence of a **universal time**. In principle such clocks can be moved to each point of physical space and the time at which a particle passes a particular point read off on that clock placed at the point in question. This procedure bypasses a difficulty which would arise if the time was read off on a clock placed at the location of an observer. Since the observer sees the particle pass the point by means of light signals, the time read off on the observer's clock will be later than the time read off on the clock placed at the location of the point. The difference in these times is equal to the time taken for the light signal to travel from the point to the observer.

Few readers would question the existence of a universal time. For example if you are arranging to meet a friend later then you might well check initially that your wristwatches show the same time but you will assume without question that subsequently your wristwatches will remain synchronized, enabling you both to arrive at your meeting place simultaneously.

As a particle moves it traces a path in physical space. In order to model this path, and the motion of the particle, it is necessary to have a mathematical model of physical space itself. The classical Newtonian mathematical model of physical space is three dimensional euclidean geometry, the path of a particle being modelled by a curve in this geometry. This mathematical model of physical space will be referred to as **euclidean space**. Notice that it is important when referring to euclidean space to remember that this is simply the geometry chosen as the mathematical model of physical space; there is no sense in which euclidean and physical space are one and the same thing. This last fact was not appreciated in previous centuries and so when the mathematician Karl Friedrich Gauss (1777–1855) found that the sum of the angles of a triangle formed by three prominent but distant landmarks added up to more than 180° he was loath to publish his conclusion that geometries exist other than the familiar geometry of Euclid. It was only in 1830, after Bolyai and Lobachevsky had published independent accounts of non-euclidean geometry, that Gauss announced his results which had been obtained thirty years earlier.

Gauss had been motivated in his work on non-euclidean geometry by his interest in the study of the shape and dimensions of the earth's surface. A student of his, Riemann, developed Gauss's ideas into a general theory and in 1919 Einstein was led to choose four dimensional Riemannian geometry as his mathematical model of space and time. Thus in Einstein's general theory of relativity, which is a theory of gravitation, physical space is no longer modelled as euclidean geometry. Relativistic mechanics is based on Einstein's earlier special theory of relativity in which space and time are modelled by a special four dimensional Riemannian geometry known as Minkowskian geometry. It is fortunate indeed that the classical model of space is three dimensional euclidean geometry, a mathematical structure with which you will have at least some familiarity.

1.5 Units and Dimensions

Many physical quantities are modelled by real numbers, the value of the real number being measured by comparing the physical quantity in question with a standard quantity of the same physical nature. The choice of the standard is indicated by specifying a **unit**. For example, lengths can be measured against several standards, two commonly used ones being the **foot** and the **metre**. Thus the length of a girder might be said to measure 15 feet or 4.6 metres, the standard length of comparison being indicated by the inclusion of the appropriate unit.

Within dynamics, certain units are taken as being **fundamental**. These are the units of length, mass and time. The system of units known as SI, that is the Systéme Internationale d'Unités, is used throughout this book, unless otherwise stated. In this system the three units are the metre, kilogram and second, respectively. These are abbreviated to m, kg and s. Other units are termed **derived units** because they can be derived from the fundamental units using the definition of the physical quantity under consideration. For example, the unit of speed is derived from the units of length and time, it is the metre per second or ms^{-1}. Similarly the unit of acceleration is the metre per second per second or ms^{-2}, the actual derivation of these two units follows from the formal definitions of speed and acceleration which are given later.

The SI unit of force is called the Newton, after Sir Isaac Newton and abbreviated to N. This unit is worth discussing in some detail. In Section 1.2 we saw, intuitively, that when a force acts continuously on a particle it is proportional to the product of the mass of the particle and the resulting acceleration of the particle. The Newton is defined to be the force which when acting on a particle of mass 1kg will result in an acceleration of $1ms^{-2}$. With this choice of unit the constant of proportionality becomes equal to one and then the force is equal to the mass of the particle times the acceleration. It is conventional to list the fundamental units in the order in which they were introduced above. Then $1N = 1m\,kg\,s^{-2}$. In the SI system certain of the derived units are named, like the Newton, after a scientist or mathematician who was intimately associated with the physical quantity in question. Another example of a derived unit which is given a specific name is the SI unit of energy, called the Joule after Sir James Joule and abbreviated by J. Other examples are listed in the following table of SI units.

Notice that the kilogram is based on the gram which is abbreviated as g. The same letter is used for this unit as is used for the gravitational acceleration. Some care has to be taken not to confuse the two!

Quantity	Unit	Abbreviation
length	metre	m
mass	kilogram	kg
time	second	s
frequency	hertz	Hz (or s^{-1})
force	newton	N (or $kgms^{-2}$)
energy	joule	J (or kgm^2s^{-2})
power	watt	W (or kgm^2s^{-3})
angle	radian	rad
electric charge	coulomb	C
electrostatic potential	volt	V
magnetic induction	tesla	T

Table 1.1 SI units

As our examples show, derived units are combinations of powers of the fundamental units of length, mass and time. As another example, area is calculated by multiplying one length by another and is measured in (unit of length)^2. Put in another way area has **dimension** 2 in length, $[L^2]$. Similarly, volume has dimension 3 in length, $[L^3]$. The square brackets are used to stand for "dimension of". For example speed has dimensions of 1 in length and -1 in time, so that

$$[\text{speed}] = [L/T] = [LT^{-1}].$$

As another example consider the kinetic energy of a particle of mass m moving with speed v. This will be defined in the next chapter to be $\frac{1}{2}mv^2$. Then

$$[\text{kinetic energy}] = [\tfrac{1}{2}mv^2] = [M(L/T)^2] = [ML^2T^{-2}].$$

Notice that the factor of a half plays no part in determining the dimensions of the kinetic energy. This is because the factor is a pure number, a **dimensionless constant**.The gravitational potential energy of a particle is also introduced in the next chapter, it is given by the formula mgh. We might expect this gravitational potential energy to have the same dimensions as kinetic energy, both quantities being energies. The mass m and height h of the particle are both variables which have dimensions. The constant g is the acceleration due to gravity and so, unlike the constant $\frac{1}{2}$ in the formula for the kinetic energy, it is not dimensionless. It has the dimensions of an acceleration, $[LT^{-2}]$, and is an example of a **dimensional constant**. The dimensions of the gravitational potential energy are therefore

$$[\text{gravitational potential energy}] = [mgh] = [M(LT^{-2})L] = [ML^2T^{-2}]$$

which are the same as those of kinetic energy.

The equality of dimensions for quantities of the same physical nature extends to all equations of mechanics; they are **dimensionally homogeneous**. That is, all the terms in a mechanics equation have the same dimensions. For example each term in the kinematics equation

$$s = ut + \tfrac{1}{2}at^2$$

for motion in a straight line with constant acceleration a has dimension 1 in length, i.e. $[L]$. The dimensional homogeneity of equations provides an invaluable check, to within dimensionless constants, of each step in the solution of any mechanics problem.

Dimensionless variables will be of importance in the next section and subsequently. Such a variable is, as its name suggests, a variable without dimensions. An example of such an animal is the angle when measured in radians. The radian will have been defined for you as the angle subtended at the centre of a circle by an arc equal in length to the radius of the circle. Similarly an angle of 2 radians will be that subtended by an arc of length equal to twice the radius of the circle, see Fig 1.4(i) and (ii). In general an angle is measured in radians by taking the ratio of the arc length which subtends the angle at the centre to the radius of the circle, see Fig 1.4(iii).

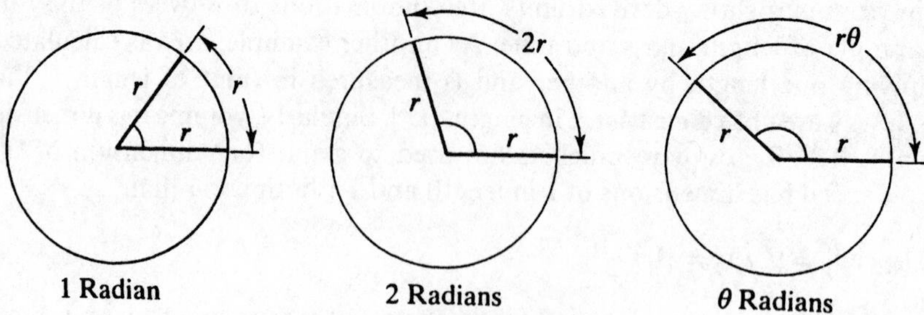

Fig 1.4 Definition of a radian.

Thus

$$\theta = \frac{\text{arc length}}{\text{radius.}}$$

Dimensionally

$$[\theta] = [L/L] = [1],$$

and hence θ has no dimensions, it is just a real number. You may know how important it is to use radians when differentiating the trigonometrical functions. There are occasions when we will find it convenient to define a dimensionless variable, particularly when using linear approximations as introduced in the next section.

TUTORIAL PROBLEM 1.7

Use the dimensional homogeneity of the equation

$$s = ut + \tfrac{1}{2}at^2$$

for motion in a straight line with constant acceleration a to discover the physical nature of the variables s and u. You should not assume any previous familiarity with the equation.

EXERCISES ON 1.5

1. A particle is moving in a circle of radius r with constant speed v. Show that

   $$[v^2/r] = [LT^{-2}].$$

   You should note that v^2/r is the only combination of the variables r and v having the dimensions of an acceleration and will therefore describe the acceleration of the particle up to a dimensionless constant.

2. A particle is swinging on the end of a string of length l, the other end being fixed. What combination of the variables l and g has dimension 1 in time? Here g is the constant acceleration due to gravity. The swinging particle describes an arc of a circle of arc length l'. Write down an analogous combination of the variables l'

and g. Does experience suggest which of the above two combinations is related to the time taken for a complete swing? If you cannot answer this last question you might experiment with an object tied onto a piece of string.

1.6 Linear Approximation as a Tool for Modelling

You may be familiar with, or at least accept, statements such as

- air pressure is a function of height
- output is a function of demand
- success is a function of effort.

These statements are somewhat vague. The first refers to two quantities, air pressure p and height h, which we know are measured as real numbers so that the functional relationship can be written as

$$p = f(h).$$

However the statement tells us nothing about the function f itself; is it a quadratic, an exponential or a trigonometrical function? Once determined, the function provides a mathematical model of the relationship between air pressure and the height at which it is measured. The second and third statements are even more vague because before we can even ask the question as to the form of the function we must first decide how output, input, success and effort are to be measured.

You will often need to model the relationship between two measurable quantities as a function, first deciding which quantity is to be chosen as the dependent variable and which as the independent variable. To find a suitable function will require the collection of data, either from observation or experiment, consisting of the values of the dependent variable measured for different values of the independent variable. The function which "best fits" this data in some specified manner is then taken as the appropriate model. One method of finding the best fit is to plot the data as points, using graph paper, and to join these points by a smooth curve. If you are lucky you may recognise the shape of this curve as being the graph of a familiar function. If this fails, you will have to resort to more sophisticated methods, perhaps involving a computer analysis of the data. Sometimes it is sufficient to specify the model in terms of the graph of the function obtained above, without specifying the actual function. Often the function arises naturally from some larger modelling process, for example as being the solution to a differential equation.

Example I

100g masses are attached to one end of a metre ruler using small amounts of blu-tack. Each time another 100g mass is added, the balance point is found and the distance, y cm of the masses from the balance point is recorded together with the total mass on the end of the ruler, x g. The experiment is illustrated below.

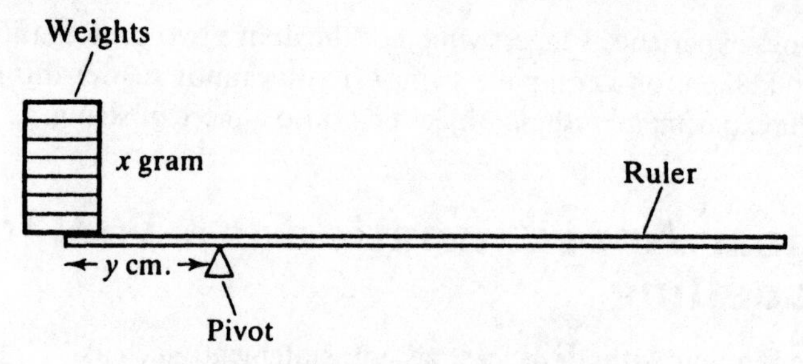

Fig 1.5 A balance point experiment.

The results of the experiment were tabulated as shown below.

Total mass on end of ruler, x.	0	100	200	300	400	500	600
Distance of masses from balance point, y.	50.0	25.5	17.5	13.0	10.5	9.0	7.5

Obtain a function describing the relationship between x and y.

SOLUTION

Plotting the given data yields the following curve.

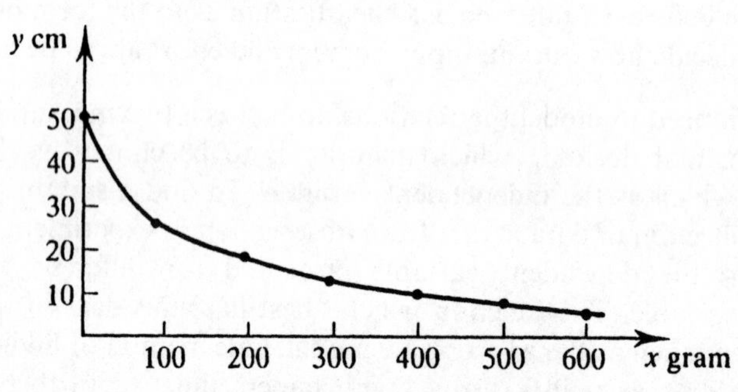

Fig 1.6 Graph of the given data.

The required function is clearly a decreasing function and the shape of the curve is suggestive of either

$$y = \frac{1}{x} \text{ or } y = e^{-x}.$$

Unfortunately the first of these functions has an asymptote at $x = 0$ whereas the curve in Fig 1.6, if it were to be extrapolated (continued) into the region $x < 0$, might have an asymptote at $x = -a$, for some suitable value of $a > 0$. This shift in the asymptote can be taken into account by considering

$$y = \frac{1}{x + a}.$$

The dimensional homogeneity of $x + a$ requires that $[a] = [\mathrm{M}]$.

The relationship

$$y = \frac{1}{x + a}$$

equates a length to the inverse of a mass and so is dimensionally inhomogeneous. It is necessary to introduce a second constant b, with $[b] = [\mathrm{ML}]$, and to consider the function

$$y = \frac{b}{x + a}.$$

Using a graphics calculator or a function graph plotting program on a computer the constants a and b can be found by trial and improvement through the plotting of functions of the above form, based upon estimates of a and b, which give a closer and closer match to the curve in Fig 1.6 obtained by plotting the original data. Alternatively we can observe that when $x = 0, y = 50$ and that given the nature of the experiment this should always be the case for a uniform meter ruler. Substituting these values into the equation gives $50 = b/a$ so that $b = 50a$. Replacing b by this expression gives

$$y = \frac{50a}{x + a}.$$

A value for a can be found by substituting a pair of values of x and y into this new equation. Substituting $x = 600\text{g}$ and $y = 7.5\text{cm}$ leads to the value $a = 106\text{g}$. Perhaps you can guess at the physical interpretation of this mass? If you know the Principal of Moments you should be able to deduce the correct interpretation. Substituting other pairs of values of x and y will lead to different values for a. If these values are close to one another then the final function

$$y = \frac{5300}{x + 106}$$

is a good model of the relationship between x and y.

The other possibility

$$y = e^{-x}$$

also needs modification in order to achieve dimensional homogeneity. Constants a and b, with $[b] = [\mathrm{M}]$ and $[a] = [\mathrm{L}]$ have to be introduced and the modified function

$$y = ae^{-\frac{x}{b}}$$

considered. The methods for finding a and b are analogous to those used above. Readers who are puzzled by the appearance of the constant b should remember that the exponential function can be defined by a power series

$$e^x = 1 + x + \frac{1}{2!}x^2 + \dots .$$

To ensure the dimensional homogeneity of the right hand side the argument of the exponential function must be chosen to be dimensionless. This is analogous to the choice of radians when discussing the trigonometrical functions. Notice also that the value of the exponential function is also dimensionless, hence the introduction of the constant a.

Perhaps the simplest function is the function

$$y = px,$$

where p is a constant, represented graphically by a straight line passing through the origin.

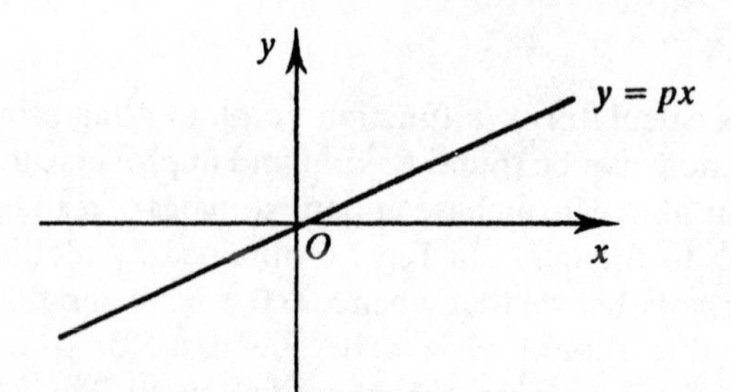

Fig 1.7 Graph of a linear function.

This function is called a linear function. You may recognize the constant p as being the slope of the straight line. Previously you will have used m to denote this slope; here the letter m will be reserved to denote the mass of a particle. A linear function is a continuous function which is differentiable for each value of x, the derivative being the slope p.

Consider now a general continuous function $y = f(x)$ whose graph, like that of a linear function, passes through the origin.

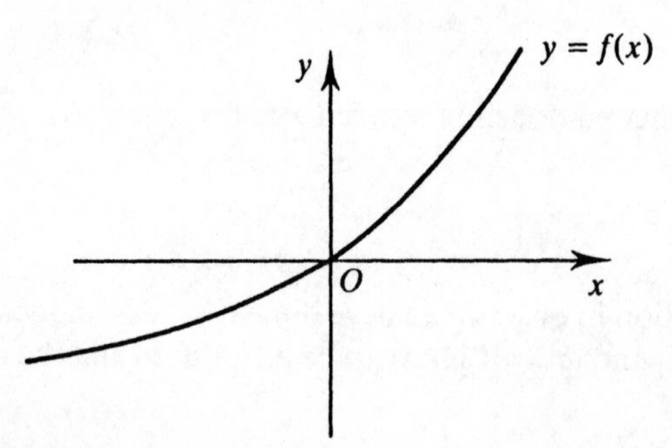

Fig 1.8 Graph of a general function passing through O.

Using a ruler add the tangent to the graph at the origin to Fig 1.8. Now place your hands at each side of the figure and slowly move them towards the y-axis. What do you notice? If you have the opportunity, use a graphical calculator or a computer package to draw the graph of any continuous function whose value at $x = 0$ is $y = 0$. Now zoom in towards $x = 0$ by reducing the scales on the x-axis. What do you notice? In each case you should observe that the curve becomes indistinguish-

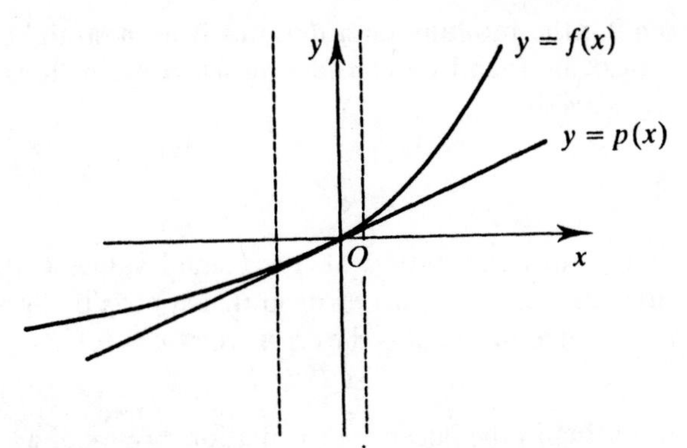

Fig 1.9 Illustrating linear approximation.

able from its tangent – perhaps you found your thumbs get in the way, in which case you could move two sheets of paper in towards the y-axis instead. In the region between the dotted lines in Fig 1.9, the difference between the curve and its tangent is less than the thickness of the drawn curve and straight line. Between these dotted lines the value of x is very small, we write this symbolically as $|x| \ll 1$; perhaps you can explain why it is important to use the modulus here? The tangent, being a straight line, represents some linear function, say $y = px$. Since for $|x| \ll 1$ the graphs of the functions $y = f(x)$ and $y = px$ are indistinguishable these functions must be approximately the same. In other words

- for $|x| \ll 1$ the function $y = f(x)$ approximates to the linear function $y = px$.

The demonstration of linear approximation given above is, of course, intuitive. A proper discussion belongs to the realm of mathematical analysis. The slope p of the tangent is equal to the slope of the graph of $y = f(x)$ at the origin, i.e. is equal to the derivative of the function $f(x)$ evaluated at $x = 0$. Thus in drawing the tangent it has been assumed implicitly that this derivative exists at $x = 0$. This is a stronger condition than continuity, and in what follows the existence of this derivative will be assumed without proof whenever we use a linear approximation.

Linear approximation is a very powerful tool, often used in mathematical modelling. A linear function can always be used to model the relationship between two measurable quantities provided that

- the relationship is described by a function whose graph passes through the origin.
- the derivative of this function exists at the origin
- attention is confined to values of the independent variable which are close to zero.

The corresponding model is called a **linear model**. This procedure bypasses the need to find a best fit function, in fact we have really demonstrated that for $|x| \ll 1$ the linear function is the best fit function. As $|x|$ gets larger the linear approximation will eventually break down. The values of x for which the linear model has to be abandoned can be found by observation or experiment, i.e. by determining where the graph representing the resulting data deviates from a straight line. Alternatively these values can be found by considering when predictions based on the linear model fail to be validated.

Example 2

The sine function $y = \sin x$ is approximated for small values of $|x|$ by a linear function. Show that when $x = 0.01$ the error in this approximation is 0.002% of the exact value of the function. What is the error when $x = 0.1$?

SOLUTION

The constant p appearing in the linear approximation $y = px$ of a function f(x) is equal to the slope of the function at $x = 0$. Now

$$\frac{d(\sin x)}{dx} = \cos x \qquad (1)$$

$$= 1 \text{ at } x = 0$$

and therefore the sine function $y = \sin x$ is approximated by the linear function $y = x$.

When $x = 0.01$ the percentage error in this approximation is given by

$$\frac{[0.01 - \sin(0.01)]}{\sin(0.01)} \times 100 = 0.002\% \quad \text{(by calculator)}.$$

When $x = 0.1$, the percentage error is

$$\frac{[0.1 - \sin(0.1)]}{\sin(0.1)} \times 100 = 0.2\% \quad \text{(by calculator)}.$$

EXERCISES ON 1.6

1. Attach a fixed mass, say one of about 100g, to one end of a piece of string which is about 150cm in length and not elastic. For different lengths of the string, l, set the mass swinging by releasing it from the same angle to the vertical on each occasion and time how long it takes the mass to do 10 complete swings. Suggested values for l are 20, 40, 60, 80, 100, 120, 140 cm. The fixed angle from which the mass is released each time should be in the region of 10 to 15 degrees.

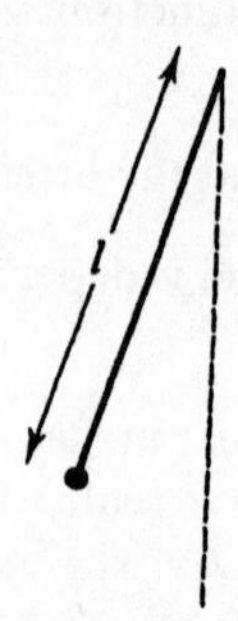

Fig 1.10 A swinging mass.

Draw a graph of the time for 10 complete swings against the length of the string in centimetres and try to fit an equation to this graph. Should you include the point (0,0) on your graph? Over what range of values of l might a linear approximation be appropriate? If you know the formula for the period of a simple pendulum in terms of its length, can you relate the constant or constants in your equation to those in the formula?

2. The function $y = (x-1)^3 + 1$ is approximated by a linear function for small values of x. Obtain this linear function using the method adopted in Example 2 and also by expanding $(x-1)^3$ in powers of x. Find the percentage error when $x = -0.01$.

1.7 Tensions and Thrusts

Different components of a mechanical system are often connected together by strings, springs or rigid rods. In this section we will consider just such a connector and discuss the forces associated with it.

TUTORIAL PROBLEM 1.8

Describe some familiar examples of connectors.

Suppose that forces of magnitude F_1 and F_2 are exerted on the end points A and B of a connector, as illustrated in Fig 1:11.

Fig 1.11 Forces exerted on a connector.

These forces are called **external** forces and are illustrated as being outwardly directed, this direction is indicated by the arrowheads. For springs and rigid rods there is no reason why the external forces should not be inwardly directed, however the situation is very different for strings. If inwardly directed external forces are applied to the ends of a string then the string will become slack, in which case the components of the mechanical system connected by the string will move independently and so the string no longer acts as a connector.

If $F_1 \neq F_2$ there is a nett force acting on the connector. The existence of such a force will cause the connector to accelerate. It follows that for a connector which is at rest, or in a state of uniform motion, $F_1 = F_2$ so that the external forces are equal in magnitude and opposite in direction. If the connector is accelerating the nett external force is proportional to the product of the mass of the connector and its acceleration. If this mass is zero then, again, the nett force must be zero so that $F_1 = F_2$. In practical situations the mass of a connector is often small compared to the masses of the components of the mechanical system to which it is attached. The connector is then described as being **light**. In the modelling process the mass of such a connector is ignored, so that the connector is considered as being **massless**.

It follows that the external forces acting on a light connector are equal in magnitude and opposite in direction, even when the light connector is accelerating.

The external force F_2 illustrated in Fig 1.11 is acting on the end point B of the connector. There is no concentration of mass at B and so the end point has zero mass (think of the end point as being an infinitesimally small segment of the connector). Using the same argument as at the end of the last paragraph, the nett force acting on B must be zero. It follows that the connector must exert an internal force on B equal in magnitude but opposite in direction to the external force. When this internal force is inwardly directed it is called the **tension** in the connector at B, when outwardly directed the word **thrust** is used. Attention will be confined throughout this book to light connectors unless the connector happens to be at rest. The tensions or thrusts at the two ends of the connector are then equal in magnitude and this magnitude will usually be denoted by T. A connector is said to be in a **state of tension/thrust** according as the internal forces acting upon its end points are tensions/thrusts or, equivalently, according as the external forces are tending to extend/compress the connector. Remember that a string can never be in a state of thrust – it will go slack before such a state is reached. T is often referred to as the tension/thrust in the connector.

Fig 1.12 Internal forces: tensions and thrusts.

You know what happens when an elastic string is pulled by the action of external forces; the string stretches. Let the length of the string in its stretched and unstretched states be l and l_0; the unstretched length l_0 is called the **natural length** of the string. The difference $x = l - l_0$ is called the **extension** of the string, it is a measure of the amount by which the string is stretched. Note that in the absence of external forces the extension x is equal to zero. Now consider what happens as the magnitudes of the external forces are increased. This last sentence is really inviting you to take part in a thought experiment in which your observations can only be based on past experience. If you wish you can perform an actual experiment by cutting an elastic band, holding each end of the resulting string and pulling; your hands are exerting the external forces. What you will find is that the larger the magnitude of the external force, the larger will be the extension. This means that the tension T in a light string, or any other elastic string which is at rest, is a continuously increasing function of the extension x, so that

$$T = f(x).$$

Furthermore the tension is zero when $x = 0$. For small extensions, $x \ll 1$, this function can be approximated by a linear function so that

$$T = px,$$

for some constant p. This linear model for the tension in an elastic string will break down as x gets larger. The critical value of x after which the model is no longer appropriate is called the **linear elastic limit** and can be found experimentally for any particular elastic string.

Unfortunately the above discussion is incomplete because the interpretation of the inequality $x \ll 1$ is ambiguous when applied to an extension x. If $x = .01$km then the inequality is certainly satisfied but the same extension, now expressed as $x = 10000$mm, no longer satisfies the inequality! The reason for this paradox is that the inequality is comparing a quantity x with physical dimensions, whose magnitude therefore depends on the choice of the unit of measurement, with a dimensionless quantity, namely 1. To overcome this difficulty we shall replace the extension x by some dimensionless variable $\bar{x}$, the obvious choice being

$$\bar{x} = \frac{x}{l_0}.$$

Being the ratio of two lengths the value of $\bar{x}$ is unchanged when the unit of length is changed. The tension T can now be written as a function of $\bar{x}$ and, in the linear model when $\bar{x} \ll 1$,

$$T = p\bar{x}.$$

The inequality $\bar{x} \ll 1$ is now unambiguous. Notice that this inequality is equivalent to $x \ll l_0$ so that the meaning of a small extension is clarified, we really mean an extension which is small compared to the natural length of the elastic string. Written in terms of x the above linear model becomes

$$T = p\frac{x}{l_0}$$

This is known as **Hooke's Law**. The constant p is usually denoted by λ and is called the **modulus of elasticity**, so that

$$T = \lambda\frac{x}{l_0}.$$

This law was first expressed as *ut tensio sic vis* by Robert Hooke (1635 – 1703), the curator of experiments to the Royal Society for the last forty years of his life and a sworn adversary of Isaac Newton.

The modulus of elasticity λ appearing in Hooke's law is a measure of the elasticity of the string. For a given external force the extension x will vary from string to string. Assuming the strings to be of equal natural lengths then the greater the extension, the more elastic is the string. For different strings, all of natural length l_0 and stretched to the same tension T, their extension x will vary inversely as the modulus of elasticity λ. Hence the more elastic the string, the smaller the modulus of elasticity. If the modulus of elasticity is very large then the extension of the string will be negligible. The string can therefore be modelled as an **inelastic** or **inextensible** string, that is a string of fixed length. The tension in an inextensible string can take any value up to a critical value beyond which the string will break. This critical value will depend on the composition of the string and can be determined experimentally. The word string used here covers not only familiar household strings and twines but also ropes, wires, chains, etc. It is unlikely that a metal chain could be considered to be light so that the previous discussion might only include such chains if they are at rest or in a state of uniform motion.

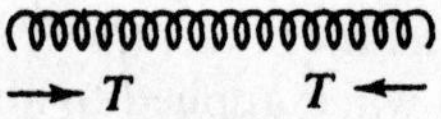

→ T T ←

Fig 1.13 A helical spring in a state of tension.

The behaviour of a spring is analogous to that of an elastic string, the only difference being that springs can be compressed as well as extended. When a spring is compressed the extension $x = l - l_0$ becomes negative. In Fig 1.13 the spring is illustrated in a state of tension. For $|x| \ll 1$,

$$T = \lambda \frac{x}{l_0}.$$

When the spring is compressed, x is negative and so T will also become negative. T is the magnitude of the tension and is a positive quantity. A negative value for T indicates that the spring is in a state of thrust rather than tension, the magnitude of the thrust being $|T|$.

The analogy between springs and elastic strings is mirrored in the analogy between rigid rods and inelastic strings. A rigid rod is of fixed length, as is an inelastic string, the only difference between the two being that rigid rods can be in a state of thrust whereas inelastic strings cannot. If a rigid rod is assumed to be in a state of tension and T is found to be negative then this indicates that the assumption is false and that the rod is actually in a state of thrust. The magnitude of the thrust is then $|T|$.

Fig 1.14 External and internal forces for a rigid rod in a state of thrust.

Example 3

A light elastic string AB consists of two portions AC and CB each of natural length l_0, their moduli of elasticity being λ and λ', respectively. The composite string is stretched to a length $3l_0$. Show that the lengths of AC and CB are in the ratio $2\lambda' + \lambda : 2\lambda + \lambda'$.

SOLUTION

Let l be the length of the portion AC when the composite string is stretched to a length $3l_0$. The tension in this portion is given by

$$T = \lambda \frac{(l - l_0)}{l_0}.$$

The length of the second portion CB is $3l_0 - l$ and so the tension in this portion is given by

$$T' = \lambda' \frac{(3l_0 - l - l_0)}{l_0} = \lambda' \frac{(2l_0 - l)}{l_0}.$$

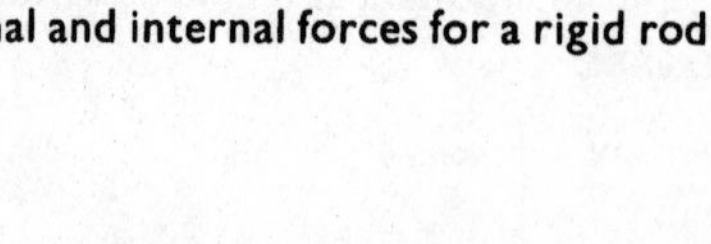

A C B

T T T' T'

Fig 1.15 A composite elastic string.

The composite string is at rest and so the nett force acting at C must be zero, i.e. $T' = T$ (the same conclusion could be drawn if the string is moving provided that no mass is attached to the string at C). Hence

$$\lambda\frac{(l-l_0)}{l_0} = \lambda'\frac{(2l_0-l)}{l_0}$$

so that

$$(\lambda+\lambda')l = (2\lambda'+\lambda)l_0.$$

It follows that the lengths of AC and CB are

$$l = \frac{(2\lambda'+\lambda)}{\lambda+\lambda'}l_0$$

and

$$3l_0 - l = \frac{(2\lambda+\lambda')}{\lambda+\lambda'}l_0$$

so that the ratio of lengths is as given.

Example 4

Three uniform elastic strings of equal natural lengths have moduli of elasticity λ_1, λ_2 and λ_3. They are joined together to form a composite string of natural length $3a$ and then stretched. Show that the tension in the composite string is as if it were a uniform string of natural length $3a$ and modulus of elasticity λ given by

$$\frac{3}{\lambda} = \frac{1}{\lambda_1} + \frac{1}{\lambda_2} + \frac{1}{\lambda_3}.$$

SOLUTION

Let the extensions of the three elastic strings be x_1, x_2 and x_3. Then the tensions in the strings are given by

$$T_1 = \lambda_1\frac{x_1}{a}, \quad T_2 = \lambda_2\frac{x_2}{a} \quad \text{and} \quad T_3 = \lambda_3\frac{x_3}{a}.$$

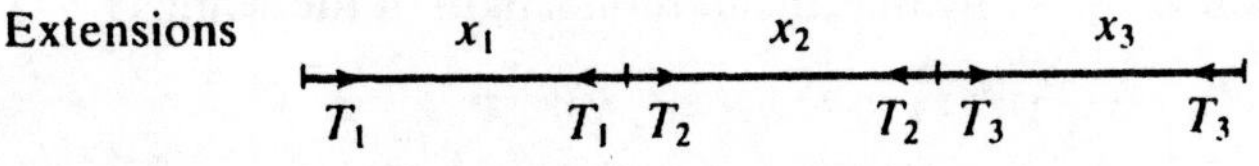

Extensions x_1 x_2 x_3

T_1 T_1 T_2 T_2 T_3 T_3

Fig 1.16 Three elastic strings joined together.

At each join the nett force is zero so that $T_1 = T_2$ and $T_2 = T_3$, i.e.

$$\lambda_1\frac{x_1}{a} = \lambda_2\frac{x_2}{a} \quad \text{and} \quad \lambda_2\frac{x_2}{a} = \lambda_3\frac{x_3}{a}.$$

It follows that

$$x_2 = \frac{\lambda_1 x_1}{\lambda_2} \quad \text{and} \quad x_3 = \frac{\lambda_1 x_1}{\lambda_3}.$$

The tension in a uniform string of natural length $3a$ and modulus of elasticity λ stretched to an extension $x_1 + x_2 + x_3$ is given by

$$T = \lambda \frac{(x_1 + x_2 + x_3)}{3a}.$$

Eliminating x_2 and x_3 this can be written

$$T = \frac{\lambda\lambda_1 x_1}{3a}\left(\frac{1}{\lambda_1} + \frac{1}{\lambda_2} + \frac{1}{\lambda_3}\right)$$

and is equal to the tension in the composite string if and only if

$$T_1 = T,$$

$$\text{i.e.} \quad \frac{\lambda_1 x_1}{a} = \frac{\lambda\lambda_1 x_1}{3a}\left(\frac{1}{\lambda_1} + \frac{1}{\lambda_2} + \frac{1}{\lambda_3}\right).$$

From this it follows that

$$\frac{3}{\lambda} = \frac{1}{\lambda_1} + \frac{1}{\lambda_2} + \frac{1}{\lambda_3}.$$

TUTORIAL PROBLEM 1.9

A non uniform elastic string of natural length l_0 has a modulus of elasticity $\lambda(x)$ which is a function of the distance x along the unstretched string from one end. The tension in this elastic string is as if it were a uniform elastic string of natural length l_0 and modulus of elasticity λ. Devise, without proof, an integral expression for $1/\lambda$.

EXERCISES ON 1.7

1. Obtain the dimensions of the modulus of elasticity from Hooke's law. Write down the SI unit of modulus of elasticity.

2. A light elastic string is of length x_1 when the tension is T_1, and of length x_2 when the tension is T_2. Show that the natural length of the string is

$$(x_2 T_1 - x_1 T_2)/(T_1 - T_2)$$

and find a corresponding expression for the modulus of elasticity. Would this same expression for the natural length hold for a light spring of length x_1 when stretched to a tension T_1, and length x_2 when compressed to a thrust T_2?

3. A light elastic string AB consists of two portions AC and CB of natural lengths a and $2a$ respectively, their moduli of elasticity being λ_1 and λ_2 respectively. The composite string is stretched to a length $4a$. Show that the length of AC is equal to

$$2a\frac{(\lambda_1 + \lambda_2)}{2\lambda_1 + \lambda_2}.$$

4. The end B of a spring AB of natural length l_0 and modulus of elasticity λ_1 is attached to one end of a rigid rod BC of length l_0. An elastic string of natural length l_0 and modulus of elasticity λ_2 connects the end A of the spring to the end C of the rod. The whole system is at rest lying along a straight line with the string taught. Show that the compressed length of the spring is given by

$$l = \frac{\lambda_1}{\lambda_1 + \lambda_2} l_0.$$

5. Show that the tension in the composite string of Example 3 is equal to the tension in a similarly stretched uniform elastic string of natural length $2l_0$ whose modulus of elasticity λ'' is given by

$$\frac{2}{\lambda''} = \frac{1}{\lambda} + \frac{1}{\lambda'}.$$

1.8 Drag Forces

A class of forces exist in nature having the perverse property of always being directed in the opposite sense to the direction of motion of the body on which they act. The direction of such a force flips whenever the direction of motion of the body changes. These forces resist motion and are called **drag forces**. Examples include the air resistance on a falling body, the drag on a vehicle, the resistance experienced when digging a spade into the ground and the force on the piston of a shock absorber as it moves through an oil filled cylinder. You may know of other examples. All such examples have a common feature, they all involve the motion of a body through some resisting medium; air, soil, oil, etc.

A very familiar drag force is the resistance of water which we have all felt when paddling in the sea whilst on holiday. You may have noticed that this resistance increases as your speed through the water increases so that the drag force is a function of the speed. This is a property common to all drag forces, the magnitude D of the drag force acting on a body moving through a resisting medium is a continuously increasing function of the speed v of the body relative to the medium so that

$$D = f(v).$$

When the speed v is zero the drag force vanishes; this is why it is easy to push a boat off from a jetty. For small speeds it is possible to approximate the function f by a linear function. However we learnt, in the last section, the importance of first introducing a dimensionless variable. We shall do this here by dividing the speed v by some other speed characteristic of the situation being modelled.

To fix ideas, consider the drag force on a sphere of diameter d moving in a fluid. The drag D will depend upon certain physical characteristics of the fluid, namely its density ρ and viscosity η (this is a measure of the "thickness" or "runniness" of the fluid). It turns out that the combination $\eta/\rho d$ has the same dimensions as speed and so

$$\bar{v} = \frac{\rho d v}{\eta}$$

is a dimensionless variable. Now

$$D = f(\bar{v})$$

and for $\bar{v} \ll 1$ this function can be approximated by a linear function to give

$$D = p\bar{v}.$$

The dimensionless variable $\bar{v}$ is called the **Reynold's number** and so for small Reynold's number the drag force is modelled as being proportional to the speed. The value of the constant p can be estimated from experiment but its form can be derived from the theory of fluid dynamics. It is found that $p = 3\pi\eta^2/\rho$ so that, written in terms of v rather than $\bar{v}$,

$$D = 3\pi\eta dv.$$

This is known as **Stokes' drag**, named after Sir George Gabriel Stokes (1819–1903). Stokes had to relinquish his Fellowship at Cambridge when he married in 1857 but was reinstated twelve years later when the university changed its statutes in order to allow married men to hold Fellowships. He became Lucasian Professor of Mathematics, the chair previously held by Newton and presently held by Stephen Hawking.

To get some idea of the conditions under which Stokes' drag is an appropriate model for the drag force on a sphere, consider first a cricket ball of diameter 7cm and mass 150g. If the ball is moving in air at normal temperature and pressure then $\rho = 1.3 \times 10^{-3}\,\mathrm{g\,cm^{-3}}$ and $\eta = 18 \times 10^{-5}\,\mathrm{g\,cm^{-1}\,s^{-1}}$. Hence

$$\bar{v} = \frac{1.3 \times 10^{-3} \times 7v}{18 \times 10^{-5}} \simeq 50v.$$

It follows that the Reynold's number $\bar{v}$ cannot be small compared to unity for any reasonable value of the speed v and so the Stokes' drag is an inappropriate model in this case. Notice that the mass of the cricket ball does not appear in the expression for $\bar{v}$ so that the same conclusion would have been reached for a tennis ball or any other ball of similar or larger dimensions. Now consider a ball bearing of radius 0.2cm dropped into a beaker of castor oil. For castor oil $\rho \simeq 1\,\mathrm{g\,cm^{-3}}$ and $\eta \simeq 15\,\mathrm{g\,cm^{-1}\,s^{-1}}$ so that

$$\bar{v} = \frac{2 \times 0.2v}{15} \simeq 0.03v.$$

In this case $\bar{v}$ will be small compared to unity provided that v is small, say less than $10\,\mathrm{cm\,s^{-1}}$. If the ball bearing is dropped into a barrel of pitch then $\rho \simeq 1\,\mathrm{g\,cm^{-3}}$ and $\eta \simeq 10^{10}\,\mathrm{g\,cm^{-1}\,s^{-1}}$. Hence

$$\bar{v} = \frac{2 \times 0.2v}{10^{10}} = 4 \times 10^{-11}v.$$

In this case $\bar{v}$ will always be small compared to unity. Even if the ball bearing is moving with the speed of light, $\bar{v}$ is less than unity! It follows from this discussion that Stokes' drag provides a very accurate model of the drag force on a ball bearing falling in pitch, a less accurate one for a ball bearing falling in castor oil and is quite inappropriate for balls used in sport

EXERCISE ON 1.8

1. Pollen grains of Picea have a radius of 70–85 microns (1 micron = 10^{-6}metres). Investigate whether Stokes' drag is an appropriate model for the drag force on this pollen as it moves in air.

Summary

- a **particle** is a physical body whose size can be neglected when modelling the particular motion of the body under consideration
- **uniform motion** is motion in a straight line with constant speed, it is the natural motion of a particle and requires no nett force to sustain it
- **particle mechanics** is the study of mathematical models describing the observed motion of a given particle (**kinematics**) and describing the relationship between the forces which are observed to disturb the uniform motion of a particle and the actual motions they cause (**dynamics**)
- classical mechanics is based on using **euclidean geometry** to model physical space and real numbers to model intervals of time; time is assumed to be **universal** so that identical clocks, once synchronized, remain synchronized no matter how they move
- speed and direction of motion are modelled together as **velocity**
- the word **force** is reserved to mean a continuously acting force
- an **impulsive force** is proportional to the mass times the resulting change in the velocity of the particle it acts upon
- **force** is proportional to the mass times the resulting acceleration of the particle it is acting on
- in each of the above the constant of proportionality can be reduced to unity by an appropriate choice of the unit of mass
- the **SI units** of measurement are based on the metre, kilogram and second as the **fundamental units** of length, mass and time
- other units are **derived** from the above using the definition of the quantity under consideration
- the equations of mechanics are **dimensionally homogeneous**
- for extensions x of an elastic string which are small compared to its natural length l_0 a linear model leads to **Hooke's law**

  $$T = \lambda \frac{x}{l_0}$$

  for the tension T in the string, λ being the **modulus of elasticity**

- Hooke's law also models the tension and thrust in a spring; when compressed the extension is negative leading to a negative value for T which is interpreted as a thrust of magnitude $|T|$
- an **inelastic** or **inextensible** string is one of constant length and is an appropriate model for strings having very large moduli of elasticity
- a **rigid rod** has constant length and, like a spring, can be in a state of tension or thrust
- a string cannot be in a state of thrust, it becomes slack before such a state is reached
- for small Reynold's number a linear model leads to **Stokes' drag**

$$D = 3\pi\eta dv$$

for the drag D on a sphere of diameter d moving with speed v in a medium of viscosity η.

2 • Straight Line Motion

Some of the intuitive ideas met in the previous chapter are developed in the context of the motion of a particle moving on a straight line. Newton's second law of motion is expressed as an equation, the general solution of which describes all possible motions of a particle consistent with the given forces acting on that particle. The importance of initial conditions to the determination of the actual motion of a given particle is stressed. The concepts of kinetic and potential energies are introduced and the conservation of the total energy used as a powerful tool for a qualitative discussion of motion. The theme of linear approximation again plays a role, particularly in the discussion of the gravitational acceleration experienced by a particle moving close to the earth's surface. The motion of a particle moving in a resisting medium is studied and the existence of a limiting speed is predicted.

2.1 Kinematics of a Particle Moving on a Straight Line

Consider a particle P moving on a straight line l as illustrated. The first step in modelling the observed motion of such a particle is to decide on a method for

Fig 2.1 Straight line motion.

specifying the position of the particle at any given instant of time. If the particle models a car in which you are travelling then you might well specify your position by observing features of the surrounding landscape – a distant spire, a roadside telephone box, intersections with other roads, etc. However, we do not usually follow this procedure when discussing straight line motion. Instead we consider the straight line in isolation rather than in relation to the rest of the space; the motion is thought of as being one dimensional. In these circumstances the position of P can only be specified in relation to some other point lying on the straight line itself. Such a point O will be referred to as an origin.

As the particle P moves on the straight line the distance of P from O will vary. This distance is not sufficient to specify the position of P relative to O; knowing the distance alone does not specify whether P is to the left or right of O. The use of the phrase to the left or right of O is itself ambiguous and depends very much on how you look at the straight line. Turning the page upside down interchanges "left of O" with "right of O". What we really require is an unambiguous method for specifying the **orientation** of P relative to O. The simplest procedure is to first orientate the line itself by placing an arrowhead on it, as has been done in Fig 2.2. The orientation of P relative to O can then be specified according as to whether a translation from O to P is in the direction of the arrowhead or not. In what follows

O P l

Fig 2.2 An origin and orientation.

the direction of the arrowhead specifying the orientation of the straight line will be called the **positive direction**, the opposite direction will be called the **negative direction.**

The distance d of P from O is a **positive** number. The position of P relative to O is specified by a **directed** number x, called the **displacement of P relative to O**, and defined by

$$|x| = d$$

with $x > 0$ if the translation from O to P is in the positive direction and $x < 0$ if the translation from O to P is in the negative direction.

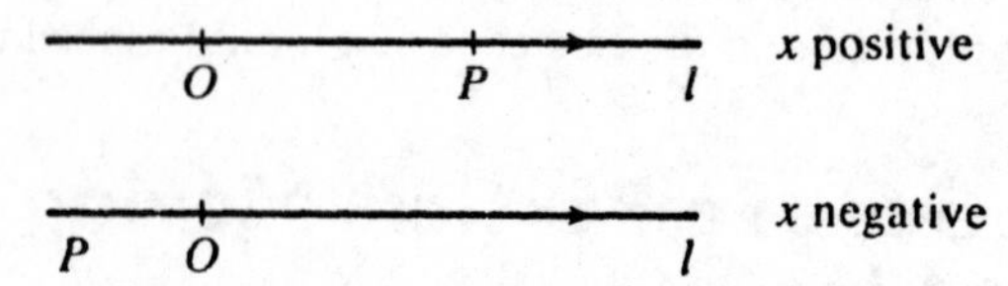

Fig 2.3 Positive and negative displacements.

Two simple but nevertheless significant results follow directly from the definition of displacement, namely

- the displacement of a point relative to itself is zero
- the displacement of O relative to P is equal to minus the displacement of P relative to O.

The actual value of the displacement x of a given point depends on the choice of orientation and on the choice of origin. Reversing the orientation changes the sign of x. To determine how the displacements of a point P relative to two different origins O and O' are related consider the three points ordered as in the following figure.

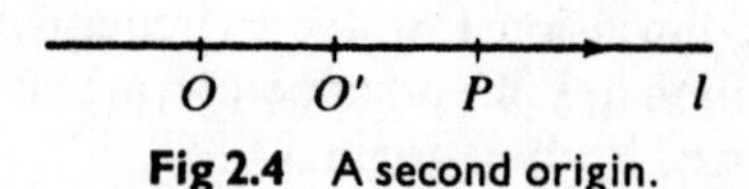

Fig 2.4 A second origin.

Now displacement of P relative to O' = distance of P from O'
= distance of P from O –
distance of O from O'
= displacement of P relative to O +
displacement of O relative to O'.

Perhaps we should abbreviate this result by writing it as

disp of P rel to O' = disp of P rel to O + disp of O rel to O'.

This result is known as the addition law of relative displacements. Notice the juxtaposition of the two O's on the right hand side, this might help you to remember the addition law.

TUTORIAL PROBLEM 2.1

Illustrate all possible ordering of three points O, O' and P lying on a straight line. Prove that the addition law of relative displacements holds in all cases.

Example

A pair of batsmen run a single. Sketch graphs of the displacements of each batsman relative to the striker's end against time. Take the orientation to be directed down the wicket from the striker's end. By considering a suitable combination of these graphs obtain the graph of the displacement of the striker relative to the non striker.

SOLUTION

Let A be the striker, B the non striker and O the striker's end. As the striker is initially stationary A will not cover much ground for the first few seconds of the

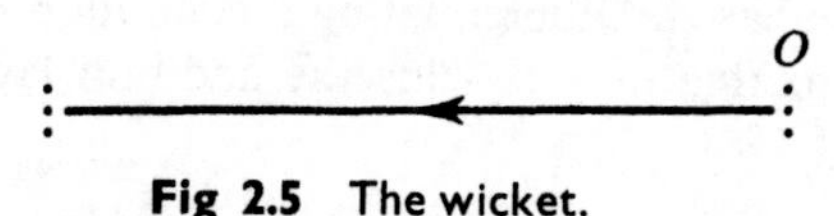

Fig 2.5 The wicket.

run and as the striker comes to the end of the wicket A will come to a stop and so again will not cover much ground for the last few seconds of the run. The same is true of B. The two graphs will therefore be as illustrated, d being the distance between wickets.

Now

$$\begin{aligned}\text{disp of } A \text{ rel to } B &= \text{disp of } A \text{ rel to } O + \text{disp of } O \text{ rel to } B\\ &= \text{disp of } A \text{ rel to } O - \text{disp of } B \text{ rel to } O.\end{aligned}$$

The required graph of the displacement of the striker relative to the non striker is therefore obtained by taking graph B from graph A, resulting in the graph C illustrated in Fig 2.6.

It is usual to consider the origin O as being the location of an observer, x is then referred to as the displacement of the particle P relative to the observer O. If the particle is moving relative to the observer its displacement x will be a function of time t,

$$x = x(t).$$

Notice here that the symbol x is being used to denote both the function and its value, a practice which would be discouraged in many courses. In a sense the usage

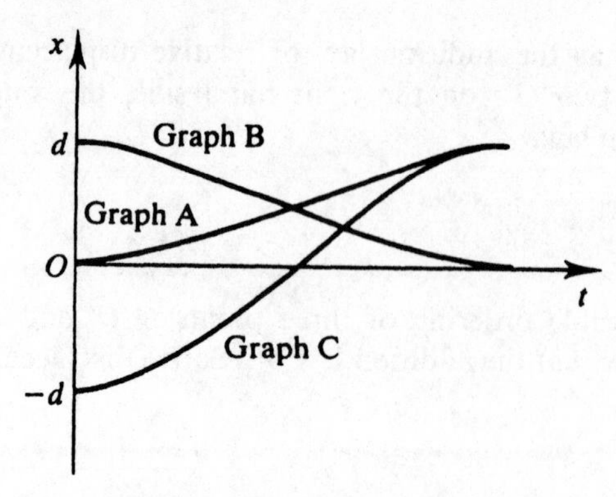

Fig 2.6 Graphs of displacement.

is a form of mathematical slang and is commonly adopted in applications. The derivative of the displacement is defined to be the **velocity of P relative to O**. This derivative is written as either $\dot{x}$ or $\frac{dx}{dt}$, the first notation being typographically convenient, the second being useful in manipulations, particularly when leading to an integration or a use of the chain rule. Choosing a different origin will, in general, result in a change of velocity. Assuming that the two observers associated with the origins O and O' are using identical clocks then, provided that these clocks are initially synchronized, both observers will see a given particle pass a given point at the same time t. Here we are using the universality of time on which all Newtonian models of motion are based. Differentiating the addition law of relative displacements with respect to this time t yields the addition law of relative velocities, abbreviated as

$$\text{vel of } P \text{ rel to } O' = \text{vel of } P \text{ rel to } O + \text{vel of } O \text{ rel to } O'.$$

Notice again the juxtaposition of the two O's on the right hand side of the equation. From the addition law it follows that the velocity of P is the same relative to both origins if and only if the velocity of O relative to O' is zero. This requires that the displacement of O relative to O' be a constant. The origin O' is then said to be at **rest relative** to O.

Notice that, like displacement, the velocity of a particle relative to a given observer is a directed number. Consider this velocity at some given instant of time. If $\dot{x} > 0$ then the displacement x is increasing with time so that the particle is moving in the positive direction. If, however, $\dot{x} < 0$ then the displacement is decreasing so that the particle is moving in the negative direction. The remaining possibility is that $\dot{x} = 0$, in which case the particle is said to be **stationary** or **instantaneously at rest** at the given instant of time. The point at which the particle is then located is called a **stationary point** of the motion. If $\dot{x} = 0$ in an interval of time rather than at a given instant then the displacement of the particle relative to the observer will be constant and the particle is then said to be **at rest**.

The concept of velocity was introduced in section 1.2 and following the discussion given there you might well expect the velocity of the particle to contain more information than just the direction of motion of the particle at each instant of time. This is indeed the case. Suppose that the particle is moving in the positive direction so that the change δx in the displacement x in a small time interval δt is positive. Then δx will measure the distance travelled by the particle in the time δt. It follows that the rate of change of distance travelled with respect to time is equal to

$$\lim_{\delta t \to 0} \frac{\delta x}{\delta t} = \dot{x}.$$

Suppose, on the other hand, that the particle is moving in the negative direction. Now δx will be negative and the distance travelled by the particle in the time interval δt will be $-\delta x$. It follows that the rate of change of distance travelled with respect to time is now equal to

$$\lim_{\delta t \to 0} -\frac{\delta x}{\delta t} = -\dot{x}.$$

Remembering that if $\dot{x} > 0$, then $|\dot{x}| = \dot{x}$ but if $\dot{x} < 0$, $|\dot{x}| = -\dot{x}$ it follows from the above that in both cases the rate of change of distance travelled with respect to time is equal to $|\dot{x}|$. In other words the modulus of the velocity is equal to the speed of the particle. Notice that the proof given here assumes that the particle is moving in one direction only, i.e. is not **oscillating**. A more general proof would involve concepts from mathematical analysis.

Example 2

A particle P moves on a straight line, its displacement relative to a given origin O at time t being vt. A second origin O' is oscillating about O, its displacement relative to O at time t being $a \sin \omega t$. Here v, a and ω are all positive constants. Derive an expression for the displacement of P relative to O' and show that the velocity of P relative to O' is equal to

$$v - a\omega \cos \omega t.$$

Deduce that if $v > a\omega$ the particle will always move in the positive direction relative to O'.

SOLUTION

Using the addition law of relative displacements,

$$\begin{aligned} \text{disp of } P \text{ rel to } O' &= \text{disp of } P \text{ rel to } O + \text{disp of } O \text{ rel to } O' \\ &= \text{disp of } P \text{ rel to } O - \text{disp of } O' \text{ rel to } O \\ &= vt - a \sin \omega t. \end{aligned}$$

Differentiating this gives

$$\text{vel of } P \text{ rel to } O' = v - a\omega \cos \omega t,$$

as required. The direction of motion of P relative to O' is determined by the sign of this velocity. Now

$$v - a\omega \cos \omega t \geq v - a\omega. \quad (\text{since } \cos \omega t \leq 1)$$

If $v > a\omega$ it follows that

$$\text{vel of } P \text{ rel to } O' > 0$$

and so P will always move in the positive direction relative to O'.

All aspects of the motion of a particle can be predicted if its displacement relative to some given origin is known as a function of time. For example, differentiating the function gives the velocity and so the particle's speed and direction of motion. The location of stationary points can be found by equating the velocity to zero, etc. etc. Alternatively all aspects of the motion of the particle can be predicted if its velocity relative to some given origin is known as a function of time, together with its displacement at some given initial instant of time. To see this let x and v denote the displacement and velocity of the particle. Then, by definition,

$$\frac{dx}{dt} = v(t).$$

This can be integrated to yield

$$x = \int v(t)\, dt.$$

You will remember that an indefinite integral includes an arbitrary constant of integration and so the displacement, and therefore the motion of the particle, is not found uniquely. As an example, if $v = \sin \omega t$ then

$$x = \int \sin \omega t\, dt.$$

This can be integrated explicitly to give

$$x = -\frac{1}{\omega} \cos \omega t + c,$$

where c is the constant of integration. Here $-\frac{1}{\omega}\cos\omega t$ is a **particular value** of the integral, adding on the constant c gives the general value of the indefinite integral. Suppose that in the general case $f(t)$ is a particular value of the integral $\int v(t)dt$. Then

$$x = f(t) + c.$$

Now the displacement, say x_0, of the particle at some initial time, say t_0, is assumed to be known and substituting these values into the above equation yields

$$x_0 = f(t_0) + c$$

so that

$$c = x_0 - f(t_0).$$

It follows that

$$x = f(t) + x_0 - f(t_0).$$

The displacement x has now been found uniquely as a function of time and so all aspects of the motion of the particle can indeed be predicted.

TUTORIAL PROBLEM 2.2

Obtain the above equation for x by integrating

$$\frac{dx}{dt} = v(t)$$

using definite integrals whose limits correspond to the initial time t_0 and the current time t.

Generalizing the result of the last paragraph, all aspects of the motion of the particle can also be predicted if some higher derivative of the displacement of the particle relative to a given origin O is known as a function of time, together with information about the motion at some given initial time which is sufficient to determine all the constants which arise when integrating to find the displacement. In fact the only higher derivative of the displacement $x(t)$ to have a physical importance is the second derivative $\ddot{x}$ or $\frac{d^2x}{dt^2}$ or even $\frac{d\dot{x}}{dt}$, to mix the two notations. This derivative is called the **acceleration** of the particle relative to the origin O. The acceleration can also be expressed as the first derivative $\frac{dv}{dt}$ or $\dot{v}$ of the velocity $v(t)$ and as such measures the rate of change of the velocity. This relates to the intuitive notion of acceleration, introduced in section 1.2, as being associated with changing speed or direction of motion, that is to changing velocity. Differentiating the addition law of relative velocities gives the addition law of relative accelerations which can be written in the abbreviated form

$$\text{acc of } P \text{ rel to } O' = \text{acc of } P \text{ rel to } O + \text{acc of } O \text{ rel to } O'.$$

TUTORIAL PROBLEM 2.3

Repeat example 1 with displacements replaced by velocities and then by accelerations.

Many pre-university courses give much attention to the motion of a particle moving with a constant acceleration. The physical importance of such motion is that it provides the classical mathematical model for the motion of a particle moving close to the earth's surface under the action of the earth's gravitational attraction. Such a particle has a constant acceleration which is usually denoted by g and called the gravitational acceleration. The value of g is approximately 9.8 ms^{-2}. This topic will be dealt with in Section 2.6.

Example 3

(i) Prove that, for straight line motion, the area under the graph of velocity v against time t is equal to the displacement of the final position of the particle relative to the initial position. Show that if v is positive this area can also be interpreted as the distance travelled.

(ii) A model for the motion of sprinters is that they accelerate uniformly from rest, achieving a particular speed which they maintain until the end of the race. Sketch a graph of velocity against time illustrating this model. In 1993, the world records for the men's 100m sprint and 60 metres dash were 9.86s and 6.41s respectively. Calculate the speed with which the races are finished, the period of time over which the acceleration occurs and the value of the acceleration.

SOLUTION

(i) The area under the graph between $t = t_1$ and $t = t_2$ is equal to the integral

$$\begin{aligned}\int_{t_1}^{t_2} v\,dt &= \int_{t_1}^{t_2} \frac{dx}{dt}dt\\ &= x(t_2) - x(t_1)\\ &= \text{disp of final position of } P \text{ rel } O - \text{disp of initial position of } P \text{ rel } O\\ &= \text{disp of final position of } P \text{ rel to initial position of } P.\end{aligned}$$

If v is positive then the displacement is increasing and so the displacement of the final position of P relative to the initial position is positive and equal to the distance travelled. In this case v can be interpreted as the speed of the particle.

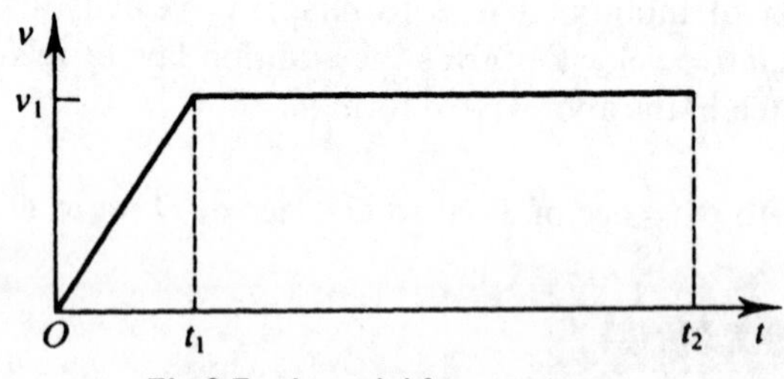

Fig 2.7 A model for a sprinter.

(ii) The required graph is as follows, where v_1 is the speed with which the race is finished and t_1 is the time over which the acceleration occurs. The area under this graph is the sum of the areas of a triangle and a rectangle so that

$$\tfrac{1}{2}v_1 t_1 + v_1(9.86 - t_1) = 100$$
$$\text{and } \tfrac{1}{2}v_1 t_1 + v_1(6.41 - t_1) = 60,$$

where we are assuming that the final speed v_1 and time t_1 are the same for both races. Subtracting these equations yield

$$v_1(9.86 - 6.41) = 40$$

so that $v_1 = 11.6\text{ms}^{-1}$. Substituting this into the first equation gives $t_1 = 2.48$s. The acceleration is the rate of change of velocity with respect to time. Here the acceleration is constant and so this rate of change is simply v_1/t_1 leading to the value 4.68 ms^{-2}.

EXERCISES ON 2.1

1. Compare the acceleration of the athlete in Example 3 to the acceleration of a Porsche which goes from 0 to 28ms^{-1} in 7 seconds.

2. A car draws away from one set of traffic lights, then slows down and stops for the next set. This pattern is then repeated for another set of traffic lights. Sketch graphs of the car's displacement, velocity and acceleration relative to the first set of traffic lights.

3. You are travelling at 50km per hour at a distance of 50m behind a lorry, travelling with the same speed, on a straight road. To overtake, you accelerate up to a speed of 65km per hour, with an acceleration of $2m$ s^{-2}, and maintain this speed, pulling in front of the lorry when 50m ahead of it. Sketch the graph of your speed relative to the lorry. From this graph, find the total time for the manoeuvre and for the period of acceleration. What length of road do you need to complete this manoeuvre?

4. (a) A train travels for 40 minutes at 60 km per hour and 20 minutes at 90 km per hour. Calculate the average speed over the hour.

   (b) A particle is moving on a straight line with a velocity given by $\dot{x} = -a\omega \sin \omega t$. Calculate the average speed between

   (i) $t = 0$ and $t = \frac{\pi}{\omega}$

   (ii) $t = 0$ and $t = \frac{2\pi}{\omega}$.

5. Careful measurements were taken from a stroboscopic photograph of a particle being uniformly accelerated. These measurements are given below, the distances are measured from the starting point but the time origin is the first clearly discernible image. Using the data, estimate the apparent acceleration of the particle in the photograph and the time which elapsed from the particle being released to the first discernible image. You may assume that the distance s travelled by the particle is given by $s = \frac{1}{2}at^2 + ut + c$, where a, u, c are constants, a being the acceleration.

Time in units of 1/30 s	0	1	2	3	4	5
Distance in cm	3.00	4.60	6.50	8.75	11.30	14.20

2.2 Dynamics of a Particle Moving on a Straight Line

In section 1.2 we saw that Newton's breakthrough in modelling motion was his realization that the force acting on a particle is proportional to the product of the mass of the particle and the resulting acceleration. This is known as Newton's second law of motion. Denoting the force by F, the mass by m and choosing the unit of force so that the constant of proportionality is equal to one, Newton's second law of motion for a particle P moving on a straight line can be written as

$$m\ddot{x} = F,$$

where x is the displacement of P relative to an observer O at rest, that is a fixed observer. The meaning of an observer being at rest or fixed will be discussed in detail later. Notice the convention used here of placing the force on the right hand side of the equation. The reason for this is that the force is usually given and the equation is then used to find the displacement $x(t)$ and so predict the motion of the particle P. The equation is known as the **equation of motion** and it presupposes that, for straight line motion, the physical concept of force can be modelled mathematically as a directed number. If F is positive then $\ddot{x} > 0$ and so the velocity $\dot{x}$ of the particle is increasing with time. If at some instant of time the particle is stationary (i.e. $\dot{x} = 0$) then subsequently $\dot{x}$ must be positive and the particle will move in the positive direction. The force F is said to be directed in the positive direction. If, however, F is negative then the particle will move from its stationary point in the negative direction. The force is then said to be directed in the negative direction. $|F|$ is called the magnitude of the force. If more than one force is acting on the particle then the force which appears in Newton's second law is the **resultant** force which is modelled mathematically by adding all the individual F's; in the previous chapter the more intuitive adjective **nett** was used.

In general, the force F acting on a given particle will be a function of the displacement x and velocity $\dot{x}$ of the particle and of the time t at which the force is measured, i.e.

$$F = F(t, x, \dot{x}).$$

If this function is known then the equation of motion $m\ddot{x} = F$ can be considered as a **differential equation** for the unknown function $x(t)$. Solving the equation will involve two integrations and so the general solution for x will contain two arbitrary constants of integration. For different values of these constants we obtain different **possible** motions of the particle consistent with the given force F (a motion is determined by a function $x(t)$). The **actual** motion of a given particle is only determined if we have at our disposal sufficient information about the initial motion of the particle to find explicit values of the constants of integration. Such information is said to constitute the **initial conditions** of motion. In many practical situations the actual form of the force $F(t, x, \dot{x})$ is so complicated that the corresponding equation of motion is too difficult to integrate directly. In such cases numerical techniques have to be used. One such technique is outlined below, based on the familiar result that the change in the value of a function due to a small change in the independent variable is approximately equal to the derivative of the function multiplied by the small change in the independent variable. This product is, in fact, the linear approximation of the actual change in the value of the function. It is often used in estimating errors. Perhaps you remember having estimated the error δA in the calculated area A of a circle due to an error δr in the measured value r of its radius, using

$$\delta A = \frac{dA}{dr}\delta r?$$

Since $A = \pi r^2$ this gives $\delta A = 2\pi r \delta r$ so that if an error of 0.01m is made in measuring the radius of a circle of radius 0.5m then the error in the calculated area of the circle is $2\pi \times 0.5 \times 0.01 = 0.03\text{m}^2$.

Consider a particle P of mass m moving on a straight line. If v denotes the velocity of the particle relative to a fixed origin O at time t then the equation of motion can be written in the form

$$m\frac{dv}{dt} = F(t, x, v).$$

Now suppose that the displacement and velocity of the particle relative to the origin O are observed at time t_0 to have the values x_0 and v_0, respectively. If δt is small then at time $t_1 = t_0 + \delta t$ the displacement of the particle will be $x_1 = x_0 + \delta x$, where

$$\delta x \approx \left.\frac{dx}{dt}\right|_{t_0} \delta t = v_0 \delta t.$$

Here $\big|_{t_0}$ denotes evaluation at $t = t_0$. At time t_1 the velocity of the particle will be $v_1 = v_0 + \delta v$, where

$$\delta v \approx \left.\frac{dv}{dt}\right|_{t_0} \delta t = \frac{1}{m}F(t_0, x_0, v_0)\delta t.$$

Having observed the displacement x_0 and velocity v_0 of the particle at time t_0 it has been possible to deduce the displacement x_1 and velocity v_1 of the particle at time t_1, a short interval δt later than t_0. Using exactly the same procedure it is possible to deduce the displacement x_2 and velocity v_2 of the particle at time t_2 a short interval δt later than t_1, or $2\delta t$ later than t_0. It can be seen that in order to find the displacement x of the particle at any subsequent time t it is only necessary to divide the time interval $t - t_0$ into n small intervals $\delta t = (t - t_0)/n$ and to apply the above procedure n times. The smaller δt is made the more accurate will be the final expression for the displacement. As δt is made small the number n of intervals becomes larger and then this numerical method for the integration of the equation of motion is only practical if carried out using a computer or programmable calculator. From this discussion it follows that the motion of a particle can be determined completely whenever the displacement x_0 and velocity v_0 of the particle at time t_0 are known. It follows that, when integrating the equations of motion directly, the values of the displacement and velocity at time t_0 are suitable initial conditions in terms of which the constants of integration can be determined.

The various forces which you will meet in this book are sufficiently simple to allow you to integrate the equation of motion directly. The simplest force is one which does not actually depend on any of t, x or v, that is a constant force. The constant gravitational acceleration g of a particle moving close to the earth's surface was introduced at the end of the last section. According to Newton's second law such a particle, of mass m, will experience a force $F = mg$. This gravitational force is called the **weight** of the particle, it is a constant force and acts vertically downwards.

Example 4

A particle of unit mass moves on a straight line, its displacement and velocity relative to a fixed point O on the line being x and v, respectively. The force acting on the particle is v^2 and $x = 0$, $v = 1$ when $t = 0$. By taking intervals of time of 0.1s, estimate the displacement and velocity of the particle after 0.5s.

SOLUTION

Here $t_0 = 0$, $x_0 = 0$, $v_0 = 1$ and $\frac{dv}{dt} = v^2$. Therefore,

$$\begin{aligned} x_1 &= x_0 + v_0\delta t & v_1 &= v_0 + \left.\tfrac{dv}{dt}\right|_{t_0}\delta t \\ &= 0 + 1 \times 0.1 & &= 1 + 1^2 \times 0.1 \\ &= 0.1 & &= 1.1 \\ x_2 &= x_1 + v_1\delta t & v_2 &= v_1 + \left.\tfrac{dv}{dt}\right|_{t_1}\delta t \\ &= 0.1 + 1.1 \times 0.1 & &= 1.1 + (1.1)^2 \times 0.1 \\ &= 0.21 & &= 1.221 \end{aligned}$$

This procedure soon becomes tedious and cumbersome. A graphics or programmable calculator with an 'Ans' function can speed up the process of calculating the values of v. Enter the first calculation for v_1 in full and press EXE. The answer, 1.1, is now located in the 'Ans' function. For each succeeding value of v, we are taking the previous answer 'Ans', and repeating the calculation. Enter this on the calculator as,

$$\text{Ans} + (\text{Ans})^2 \times 0.1$$

and press the execute button repeatedly. Each press will give the next value of v, beginning with the value of v_2.

$$\begin{aligned} v_2 &= 1.221 & &\text{when } t = 0.2 \\ v_3 &= 1.370 & &\text{when } t = 0.3 \\ v_4 &= 1.558 & &\text{when } t = 0.4 \\ v_5 &= 1.800 & &\text{when } t = 0.5 \end{aligned}$$

Unfortunately, the calculation of x must proceed as before unless some programming is undertaken. Thus

$$\begin{aligned} x_3 &= 0.21 + 1.221 \times 0.1 = 0.3321 \\ x_4 &= 0.3321 + 1.370 \times 0.1 = 0.4691 \\ x_5 &= 0.4691 + 1.558 \times 0.1 = 0.6249 \end{aligned}$$

For this particular example the displacement x and velocity v at time t can be obtained by direct integration. However it is worth bearing in mind that many differential equations cannot be solved by direct integration and in such cases, numerical methods such as that used above have to be employed. Integrating

$$\frac{dv}{dt} = v^2$$

in the form

$$\int \frac{1}{v^2}\,dv = \int dt$$

gives

$$-\frac{1}{v} = t + c.$$

The constant of integration c can be evaluated by substituting $t_0 = 0$ and $v_0 = 1$ to give $c = 1$. Hence

$$v = \frac{1}{1-t}.$$

From this

$$\int dx = \int \frac{dt}{1-t}$$

so that

$$x = -\log(1-t) + c'.$$

Substituting $t_0 = 0$ and $x_0 = 0$ gives $c' = 0$ and therefore

$$x = -\log(1-t).$$

Using the above it follows that, when $t = 0.5$,

$$x = 0.69 \quad \text{and} \quad v = 2.$$

The numerical solution is at variance with the exact solution. A better estimate to the exact solution can be obtained by taking a smaller interval of time than 0.1s. However more iterations would then be required.

EXERCISES ON 2.2

1. A cylindrical bottle which has water in it to a height of 10cm has a small hole in the base through which the water slowly leaks out. The rate of change of the height of the water, x cm, is given by $-x^{1/2}$. Using numerical techniques, estimate, to the nearest second, the time for the bottle to empty.

2. A colony of bacteria increases at a rate which is equal to half the number of bacteria present. At time $t = 0$, the number of bacteria present is 100. Using numerical techniques find how long it takes for the number of bacteria to double, to the nearest tenth of a second.

3. A particle is moving on a straight line and its displacement at time t, relative to a fixed point O on the line, is x. When $t = 0$, the particle passes O moving with a speed 10ms^{-1} in the positive direction. If the acceleration of the particle is given by $-v$, where v is the velocity of the particle, estimate using numerical techniques when and where the particle is first travelling with half its initial speed.

4. Compare your answers to questions 1–3 with those obtained by direct integration.

2.3 Motion Under a Time Dependent Force

Consider a particle of mass m moving on a straight line under the action of a force F which depends only on the time t at which the force is measured. The equation of motion becomes

$$m\ddot{x} = F(t).$$

To integrate this equation you will find it convenient to write it in the form

$$\frac{d\dot{x}}{dt} = \frac{F(t)}{m}.$$

Integrating with respect to time gives

$$\dot{x} = \int \frac{F(t)}{m} dt,$$

$$\text{or } \dot{x} = f(t) + c_1,$$

where $f(t)$ is a particular value of the integral and c_1 is a constant of integration. Rewriting this equation in the form

$$\frac{dx}{dt} = f(t) + c_1$$

a second integration with respect to time gives

$$x = \int f(t)dt + c_1 t$$

$$\text{or } x = g(t) + c_1 t + c_2$$

where $g(t)$ is a particular value of the integral $\int f(t)dt$ and c_2 is a second constant of integration. With experience it becomes unnecessary to rewrite derivatives explicitly in the form $\frac{d}{dt}$ before integrating with respect to time.

The displacement of the particle at time t has been found very easily in this case. The function $x(t)$ includes two arbitrary constants and describes all possible motions of the particle. To find the actual motion of the particle we require some initial conditions. Suppose that at time $t = 0$ the particle is observed to pass the point x_0 with velocity v_0. Here "the point x_0" means the point with displacement x_0 relative to the given origin O. Substituting this information into the equations obtained above for $\dot{x}$ and x gives

$$v_0 = f(0) + c_1$$

$$\text{and } x_0 = g(0) + c_2.$$

These two equations determine the constants c_1 and c_2 uniquely. With hindsight it would have been rather clever to choose the particular values $f(t)$ and $g(t)$ of the integrals to satisfy $f(0) = 0$ and $g(0) = 0$, the constants of integration could then have been interpreted directly as the initial velocity and initial displacement. This can be done by evaluating the two definite integrals

$$f(t) = \int_0^t \frac{F(t)}{m} dt \text{ and } g(t) = \int_0^t f(t)dt.$$

It is very important to understand that there is no unique choice for the initial conditions. This can be seen quite easily by considering whether the above constants of integration can be determined without knowing both the initial displacement and velocity of the particle explicitly. Suppose that at time $t = 0$ the particle is observed to pass the point x_0 and that at a later time $t = t_1$ the particle is observed to pass the point x_1. Substituting this information into the equation for x gives

$$x_0 = g(0) + c_2$$
$$\text{and } x_1 = g(t_1) + c_1 t_1 + c_2.$$

These two equations can be solved algebraically to give c_1 and c_2 and hence to determine the actual motion of the particle uniquely. The common property of all the different possible initial conditions is that from each the initial displacement and velocity can be determined either directly or by using one or both of the integrals of the equation of motion.

Example 5

A particle of mass m moves on a straight line under the action of a force $F(t) = F_0 \sin pt$, where F_0 and p are positive constants. The velocity of the particle at time $t = 0$, relative to a fixed origin O, is v_0. Show that the particle will move towards $x = +\infty$ if and only if $v_0 > -F_0/mp$. Describe the motion of the particle when $v_0 = -F_0/mp$.

SOLUTION

The equation of motion

$$m\ddot{x} = F_0 \sin pt$$

can be integrated twice to give

$$m\dot{x} = -\frac{F_0}{p}\cos pt + c_1$$
$$\text{and } mx = -\frac{F_0}{p^2}\sin pt + c_1 t + c_2.$$

The term $-\frac{F_0}{p^2}\sin pt$ in the expression for the displacement of the particle is bounded, in fact it represents an oscillation. As t increases this term will become negligible compared to $c_1 t$ and so the particle will move towards $x = +\infty$ if and only if $c_1 > 0$. Now $\dot{x} = v_0$ when $t = 0$ so that

$$mv_0 = -\frac{F_0}{p} + c_1$$

and the inequality $c_1 > 0$ can be written as

$$mv_0 + \frac{F_0}{p} > 0$$

from which $v_0 > -\frac{F_0}{mp}$. Similarly the particle will move towards $x = -\infty$ if and only if $c_1 < 0$ which leads to

$$v_0 < -\frac{F_0}{mp}.$$

When $v_0 = -\frac{F_0}{mp}$ the constant of integration c_1 becomes zero so that

$$mx = -\frac{F_0}{p^2}\sin pt + c_2$$

and the motion is purely oscillatory.

EXERCISES ON 2.3

1. The particle of Example 5 is observed to pass the points x_0 and x_1 at times $t = 0$ and $t = 2\pi/p$. Find alternative conditions for the different motions discussed in the example in terms of x_1 and x_2.

2. A particle of mass m is moving on a straight line under the action of an exponentially decreasing force $F = F_0 e^{-\lambda t}$, where F_0 and λ are positive constants. The particle passes the point x_0 with velocity v_0 at time $t = 0$. Obtain an expression for the displacement of the particle at time t and sketch a graph of the displacement x against time for the special case when $v_0 = -F_0/m\lambda$.

2.4 Motion Under a Displacement Dependent Force

Consider a particle of mass m moving on a straight line under the action of a force F which depends only on the displacement x of the particle relative to a given fixed origin O. The equation of motion becomes

$$m\ddot{x} = F(x).$$

Because of the form of the right hand side it would seem natural to integrate this equation with respect to x rather than t. For any function f,

$$\int f dx = \int f \frac{dx}{dt} dt$$

and so integrating with respect to x is equivalent to multiplying by the velocity and integrating with respect to t. Applying this procedure to the equation of motion gives

$$\int m\ddot{x}\dot{x} dt = \int F\dot{x} dt$$

$$\text{or} \quad \int m\frac{d\dot{x}}{dt}\dot{x} dt = \int F\frac{dx}{dt} dt$$

$$\text{i.e.} \quad \int m\dot{x} d\dot{x} = \int F dx.$$

The left hand side of this equation can be integrated trivially. In order to integrate the right hand side a function $V(x)$ is defined by the equation

$$F(x) = -\frac{dV}{dx}.$$

The equation now becomes

$$\int m\dot{x}d\dot{x} = -\int \frac{dV}{dx}dx$$

which is integrated to yield

$$\tfrac{1}{2}m\dot{x}^2 + V(x) = \text{constant}.$$

The quantity $\frac{1}{2}m\dot{x}^2$, that is half the mass times the square of the speed, is defined to be the **kinetic energy** of the particle and is denoted by T. $V(x)$ is defined to be the **potential energy** of the particle or the **potential** of the force $F(x)$. The sum of these two quantities is called the total energy and will be denoted by E. Hence

$$\tfrac{1}{2}m\dot{x}^2 + V(x) = E$$

where, according to the previous equation, E is a constant. This equation expresses the conservation of the total energy of the particle; it is usually referred to as **the energy equation**. You should note the method by which this equation has been obtained. Multiplying by the velocity and integrating with respect to time generalizes to the three dimensional motion of a particle, to the n dimensional space associated with the analytic dynamics of a system and to the infinite dimensional motion of a continuous distribution of mass!

The differential equation $F(x) = -dV/dx$ only defines the potential energy $V(x)$ up to an arbitrary constant of integration. This equation can be integrated and the potential energy expressed as a definite integral in the form

$$V(x) = -\int_{x_0}^{x} F(x)dx.$$

The arbitrariness in the definition of the potential energy is now manifest in the arbitrariness in the choice of the lower limit x_0. Notice that $V(x_0) = 0$ so that making a choice of x_0 fixes the position of zero potential energy. Changing x_0 merely adds a constant onto the potential energy. The integral $\int_{x_0}^{x} F(x)dx$ is defined to be the **work done by the force** in moving the particle from the point x_0 to the current point x. The integral $-\int_{x_0}^{x} F(x)dx$ is defined to be the **work done against the force**. Hence the potential energy is the work done against the force in moving the particle from a given point x_0 to the current point x.

The SI unit of energy is the joule (J) and it follows from above that one joule is the work done against a constant force of one newton in moving a particle through one metre in the direction of the force (J = Nm).

One further definition can usefully be added to those already made in this section. The **power** supplied by a force is defined to be the rate of change of the work done by the force with respect to time. Hence

$$\text{Power} = \frac{d}{dt}\int_{x_0}^{x} F(x)dx = -\frac{d}{dt}V(x).$$

Using the energy equation in the symbolic form $T + V = E$ it follows also that

$$\text{Power} = \frac{d}{dt}T(x).$$

Hence the power is the rate of change of the kinetic energy with respect to time. The SI unit of power is the watt (W) and is equal to a rate of one unit of work per second ($W = Js^{-1}$).

Example 6

One end A of an elastic string AB is fixed and the string is stretched by pulling the other end B in the direction of the string. Assuming that Hooke's law applies, show that the potential of the tension T acting at the end B is equal to $\lambda x^2/2l_0$, up to an additive constant, where x is the extension, λ the modulus of elasticity and l_0 the natural length of the string.

SOLUTION

Suppose that O is the position of the end B of the elastic string when in its unstretched state.

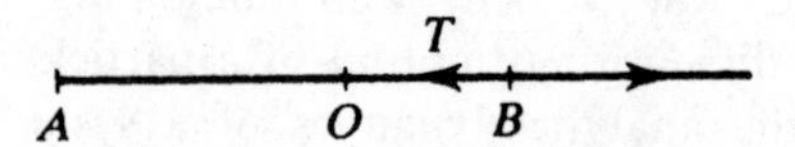

Fig 2.8 The stretched string.

The extension x of the string then measures the displacement of the end B relative to O when the string is stretched. The orientation of the tension T acting at B and of the straight line along which B moves are indicated in the figure. According to Hooke's law

$$T = \lambda\frac{x}{l_0}$$

and so, taking into account the orientation of the tension, the corresponding force F is given by

$$F = -\lambda\frac{x}{l_0}.$$

The potential V of this force is found from the equation

$$\frac{dV}{dx} = -F = \lambda\frac{x}{l_0}$$

so that

$$V = \int \lambda\frac{x}{l_0}dx = \frac{\lambda x^2}{2l_0} + \text{constant},$$

as required. The potential is usually chosen to be zero when the string is unstretched, that is when $x = 0$. The constant of integration then reduces to zero.

TUTORIAL PROBLEM 2.4

Discuss what expression is appropriate for the sum of the potentials of the tensions acting at the two ends of an elastic string, assuming that each end moves on a straight line containing the string. This sum is often referred to as the **potential energy stored in the string.**

The energy equation $\frac{1}{2}m\dot{x}^2 + V(x) = E$ is a first integral of the equation of motion. A second integration can be carried out by rewriting the equation as

$$\frac{dx}{dt} = \pm\sqrt{2(E - V(x))/m},$$

the ambiguity in sign arising on taking the square root. Integrating this equation yields

$$\int \frac{dx}{\sqrt{2(E - V(x))/m}} = \pm t.$$

In principle, knowing $V(x)$, the integral can be evaluated. This introduces a constant of integration c and the resulting equation can be inverted to give the displacement of the particle as a function of time, $x(t)$. Two constants of integration E and c will appear in this function which therefore describes all possible motions of the particle consistent with the given force $F(x)$. As in the last section initial conditions are required in order to determine the two constants of integration and therefore the actual motion of a given particle. In practice the ambiguity in the sign, the evaluation of the integral for all but the simplest potential functions $V(x)$ and the inversion of the final equation can all cause great difficulties. It is perhaps fortunate therefore that many aspects of a particle's motion can be analyzed using the energy equation without having to find x explicitly as a function of t. This method of analysis will be discussed in the next section.

Example 7

Solve the equation $F = -dV/dx$ to find the potential V of the force $F = \lambda/x^3$. A particle P of mass m is moving on a straight line under the action of the above force F. Write down an equation expressing the conservation of energy of the particle. The particle passes the point $x = a$ with velocity u and the point $x = 2a$ with velocity $2u$. Prove that the constant λ is given by

$$\lambda = 4ma^2u^2.$$

With what speed does the particle pass the point $x = 3a$?

SOLUTION

Integrating

$$\frac{dV}{dx} = -\frac{\lambda}{x^3}$$

yields

$$V = \frac{\lambda}{2x^2} + \text{constant.}$$

The energy equation is therefore

$$\tfrac{1}{2}m\dot{x}^2 + \frac{\lambda}{2x^2} = E.$$

The potential of the force is only determined up to an arbitrary constant. When writing down the energy equation this constant can always be absorbed into the constant E. Here this is equivalent to defining the potential to be zero at infinity, as can be seen by substituting $V = 0$ and $x = \infty$ in the expression for V. Substituting the values $x = a$, $\dot{x} = u$ and $x = 2a$, $\dot{x} = 2u$ into the energy equation gives

$$\tfrac{1}{2}mu^2 + \frac{\lambda}{2a^2} = E$$

$$\text{and} \quad \tfrac{1}{2}m4u^2 + \frac{\lambda}{8a^2} = E.$$

Eliminating E between these equations yields

$$\tfrac{1}{2}mu^2 + \frac{\lambda}{2a^2} = \tfrac{1}{2}m4u^2 + \frac{\lambda}{8a^2}$$

from which it follows that

$$\lambda = 4ma^2u^2.$$

Substituting this into either equation yields

$$E = \frac{5mu^2}{2}.$$

The energy equation can now be written in the form

$$\tfrac{1}{2}m\dot{x}^2 + \frac{2ma^2u^2}{x^2} = \frac{5mu^2}{2}.$$

Putting $x = 3a$ the corresponding velocity $\dot{x}$ satisfies

$$\tfrac{1}{2}m\dot{x}^2 + \frac{2ma^2u^2}{9a^2} = \frac{5mu^2}{2}$$

so that

$$\dot{x}^2 = \frac{41}{9}u^2.$$

Hence

$$\dot{x} = \pm\frac{\sqrt{41}}{3}u.$$

Notice that the sign cannot be determined so that we cannot find the velocity $\dot{x}$ of the particle as it passes the point $x = 3a$. However the corresponding speed $|\dot{x}|$ can be found, it is

$$\frac{\sqrt{41}}{3}u.$$

EXERCISES ON 2.4

1. The kinetic energy T of a particle is plotted against the displacement x. Show that the slope of the graph is equal to the force acting on the particle.

2. Find the potential of the forces

$$\text{(i) } F(x) = \lambda x \qquad \text{(ii) } F(x) = \frac{\lambda}{x^2}.$$

3. A particle of mass m moves on a straight line under the action of a force $F(x) = -kx^2$, where k is a positive constant. Write down the energy equation. At the point $x = 0$ the particle is moving in the positive x direction with speed u. The particle subsequently passes the point x_0 with speed $2u$. Show that this point is a distance

$$\sqrt[3]{\frac{9mu^2}{2k}}$$

from the origin. What can be said about the motion of the particle before it reaches this point?

4. Show that the change in the potential energy stored in an elastic string in increasing its extension is equal to the increase in extension multiplied by the average of the tensions before and after the increase.

2.5 Qualitative Discussion of Motion Based on the Energy Equation

In this section we will be concerned with the motion of a particle moving on a straight line with potential energy $V(x)$. Many of the features of the motion can be deduced from the graph of the potential energy. The slope of the graph, dV/dx, is equal to minus the force acting on the particle. Several simple deductions follow from this fact:

- the slope of the graph is positive if and only if the force is negative; for motion confined to the region $x \geq 0$ the force is directed towards the origin (such a force is called **attractive**)
- the slope of the graph is negative if and only if the force is positive; for motion confined to the region $x \geq 0$ the force is directed away from the origin (such a force is called **repulsive**)
- the slope of the graph is zero if and only if the force and therefore the acceleration of the particle is zero at the corresponding stationary point of $V(x)$

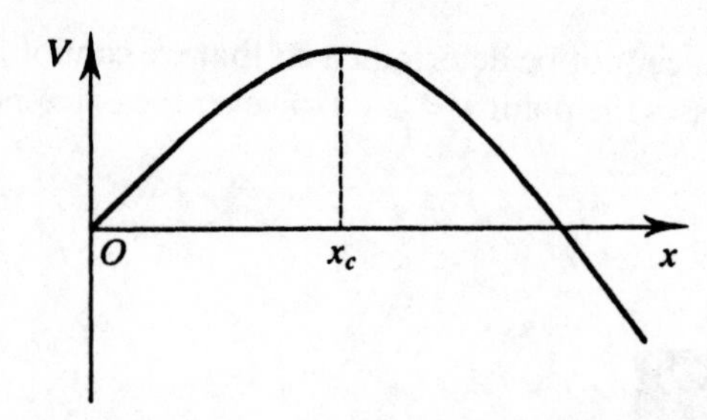

Fig 2.9 The potential of a force which changes character at $x = x_c$.

For the particular potential energy whose graph is illustrated in Fig 2.9, the force is attractive at points in the region $x < x_c$ and repulsive in the region $x > x_c$.

Gravitational forces provide examples of forces which are attractive at all points.

A full quantitive discussion of the motion of a particle moving with potential energy $V(x)$ depends on a second integration of the equation of motion as in the last section; here we will confine attention to a qualitative discussion of the most significant features of the motion. The method depends on the simple observation that the kinetic energy $\frac{1}{2}m\dot{x}^2 \geq 0$. It follows from the energy equation $\frac{1}{2}m\dot{x}^2 + V(x) = E$ that $V(x) \leq E$ and so the particle cannot be located at points x for which the graph of $V(x)$ lies above the line $V(x) = E$. For the sake of argument suppose that the graph of the potential energy $V(x)$ is as illustrated in Fig 2.10 and that the total energy E of the particle is such that the line $V(x) = E$ is located as shown in the figure.

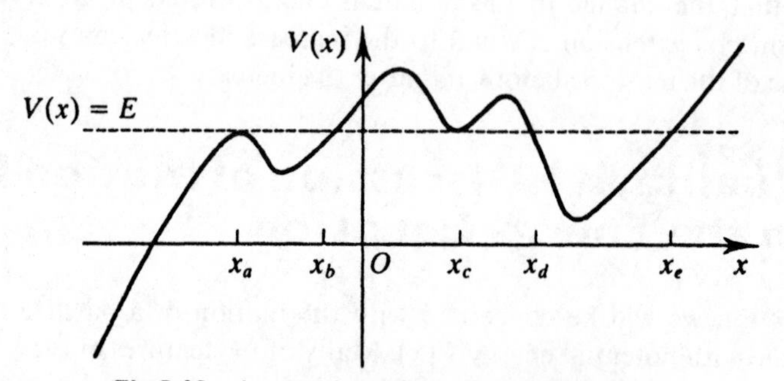

Fig 2.10 A typical graph of $V(x)$ against x.

From this graph it follows that the particle cannot be located at points x lying in the regions $x_b < x < x_c$, $x_c < x < x_d$ and $x > x_e$. Suppose that the particle lies in the region $x_d < x < x_e$ and is moving in the positive x direction. The particle will continue to move until its velocity $\dot{x}$ becomes zero when the particle will come to rest, either instantaneously or permanently. From the energy equation, $\dot{x} = 0$ if and only if $V(x) = E$. Thus the particle will approach the point $x = x_e$. At this point $V(x) = E$ and the velocity of the particle becomes zero but, because the point is not a stationary point of $V(x)$, the acceleration of the particle is non zero. This means that the particle comes to rest instantaneously at $x = x_e$. The particle cannot move into the region $x > x_e$ and so the only possibility is that it subsequently moves away from the point $x = x_e$ in the negative x direction. It will then approach $x = x_d$ and, by an argument analogous to that given above, at $x = x_d$ the direction of motion of the particle will change and it will once again move in the positive x direction, approaching $x = x_e$. The points $x = x_e$ and $x = x_d$ are called **turning points** of the motion and the particle will oscillate between these two turning points. A similar situation will occur if the particle lies in the region $x_a < x < x_b$ and is moving in the positive x direction. The particle approaches $x = x_b$ and at that point its direction of motion changes. It will then approach $x = x_a$. The difference now is that the potential energy is stationary at $x = x_a$ and so if the particle actually reaches this point it will have zero velocity and acceleration and so will remain permanently at rest at this point. There will be no change in direction of motion at $x = x_a$ and therefore the motion will not be oscillatory in this case. In practice the particle will not actually reach the point, it will take an infinite time to do so. If the particle is moving in the negative x direction in the region $x < x_a$ then it will move off to $x = -\infty$; we say that the particle **escapes to minus infinity**.

The points $x = x_a$ and $x = x_c$ merit special attention. At both points the potential energy $V(x)$ is stationary and therefore the force acting at either point is zero. Hence if a particle is placed at rest at either of these points it will remain at rest. Such points are called **positions of equilibrium** of the particle. Now suppose that the particle is disturbed slightly from its state of rest at either point. This is equivalent to adding a very small amount of energy onto the constant E. The new value of the total energy will lead to a new line $V(x) = E$ lying just "above" the original line. Imagine this new line drawn on Fig 2.10. The graph close to $x = x_a$ lies wholly below the new line so that the particle will move away from $x = x_a$. In contrast only a small section of the graph close to $x = x_c$ lies below the new line so that the particle will oscillate in the neighbourhood of $x = x_c$. There two different kinds of behaviour are reflected in the definitions of positions of **stable** and of **unstable** equilibrium as being points at which the potential energy $V(x)$ is a minimum and a maximum, respectively. A particle is said to be in stable or unstable equilibrium if it is located at rest in a position of stable or unstable equilibrium, respectively.

You have seen above that much information concerning the motion of a particle moving on a straight line under the action of a force depending on displacement alone can be obtained from the graph of the potential energy $V(x)$. This method will be met several times in later chapters. In order to utilize the method it is important to know the total energy E of the particle. In a particular problem this quantity is either given or can be found by substituting the initial conditions into the energy equation.

TUTORIAL PROBLEM 2.5

A particle moves on a straight line under the action of a force with potential $V(x)$ whose graph is illustrated in Fig 2.11. The particle is observed to move in from $-\infty$ with total energy E. If $E < V_a$ the particle will eventually change

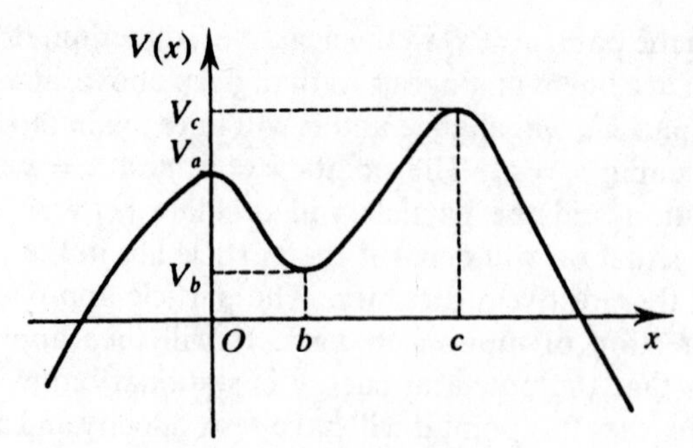

Fig 2.11 Tutorial problem.

its direction of motion and return towards $-\infty$. If $E = V_a$ the particle will approach the origin without ever reaching it. What happens if $E > V_a$?

Example 8

The expression

$$V(x) = \frac{c_1}{x^{12}} - \frac{c_2}{x^6}$$

where c_1 and c_2 are positive constants is an appropriate mathematical model for the potential of the force between two atoms in a diatomic molecule, x being the distance between the atoms. For a molecule in which one atom is very much more massive than the second atom it may be assumed that the first atom is fixed at the origin whilst the second atom moves on a straight line. Describe the possible motions of the second atom and also find its position of stable equilibrium.

SOLUTION

The atoms cannot pass through each other and so the displacement of the second relative to the first can be taken to be positive and equal to the distance x. In order to sketch the graph of $V(x)$, note the following,

- as $x \to \infty$, $V(x) \to -0$
- as $x \to 0$, $V(x) \to \infty$
- $V(x) = 0$ when $x = a$, where $a = \sqrt[6]{c_1/c_2}$
- $\frac{dV}{dx} = 0$ when $\frac{-12c_1}{x^{13}} + \frac{6c_2}{x^7} = 0$, i.e. when $x = \sqrt[6]{2}a$

Using the above the graph opposite is obtained.

The required position of stable equilibrium is the point $x = \sqrt[6]{2}a$ at which the potential is a minimum. If the total energy E is equal to the minimum value of $V(x)$, $V_{\min}$, the second atom will remain at rest relative to the first at this point. If $V_{\min} < E < 0$ the second atom will oscillate between two values of x, corresponding to the intersections of the line $V(x) = E$ and the graph. Finally if $E \geq 0$ the second atom will escape to infinity, possibly after one change of direction.

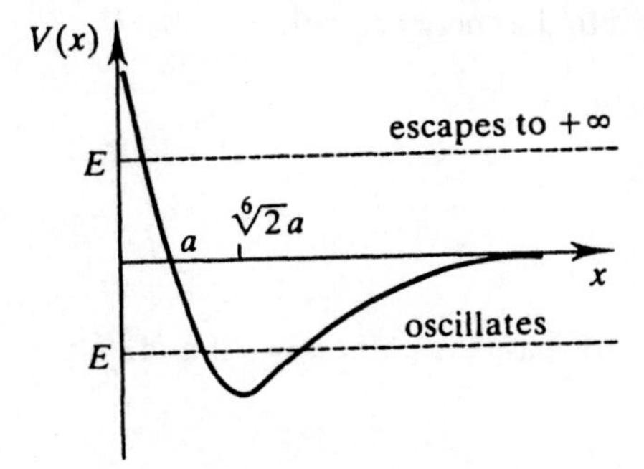

Fig 2.12 The diatomic molecule.

Example 9

A particle of mass m moves on a straight line under a force having potential energy

$$V(x) = \frac{\lambda x^3}{x^4 + a^4},$$

where λ and a are positive constants. Sketch the graph of $V(x)$ against x. The particle passes the origin, moving in the positive x direction with speed v. Prove that the particle will subsequently pass the point $x = a$ if and only if

$$v^2 > \frac{\lambda}{ma}.$$

Find a further condition on v^2 that the particle subsequently passes the point $x = -a$.

SOLUTION

In order to sketch the graph of $V(x)$ against x, note the following

- as $x \to \pm\infty$, $V(x) \to \pm 0$
- $V(x) = 0$ when $x = 0$
- $\frac{dV}{dx} = 0$ when $(x^4 + a^4)3x^2 - 4x^3x^3 = 0$, i.e. when $x = 0$ or $x = \pm\sqrt[4]{3}a$.

Using the above the following graph is obtained.

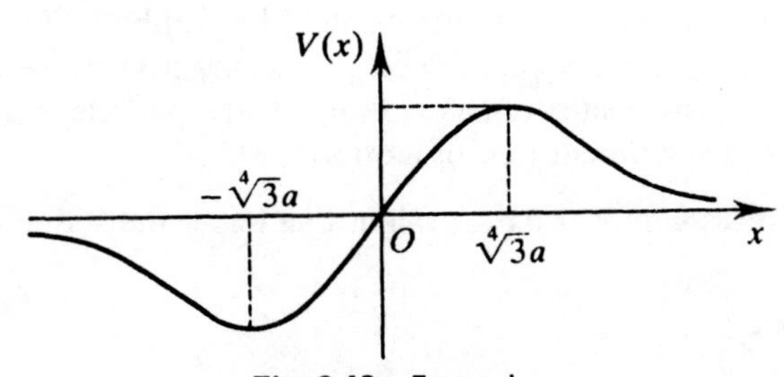

Fig. 2.13 Example.

Substituting $x = 0, \dot{x} = v$ into the energy equation

$$\tfrac{1}{2}m\dot{x}^2 + \frac{\lambda x^3}{x^4 + a^4} = E$$

yields

$$E = \tfrac{1}{2}mv^2.$$

The particle will subsequently pass the point $x = a$ if and only if

$$E > V(a)$$

$$\text{i.e.}\quad \tfrac{1}{2}mv^2 > \frac{\lambda a^3}{2a^4}$$

$$\text{or}\quad v^2 > \frac{\lambda}{ma}.$$

The particle will subsequently pass the point $x = -a$ if and only if its direction of motion reverses, i.e. if and only if

$$E < V_{\text{max}}.$$

Now

$$V_{\text{max}} = \frac{\lambda\sqrt[4]{27}a^3}{3a^4 + a^4}$$

and so the required condition is that

$$v^2 < \frac{\sqrt[4]{27}\lambda}{2ma}.$$

EXERCISES ON 2.5

1. A particle moves on a straight line under the action of a force with potential

$$V(x) = (x - 2a)^2(x + a),$$

where a is a positive constant. Sketch the graph of $V(x)$. The particle is observed passing the origin and travelling in the positive x direction. Use your graph to discuss the subsequent motion of the particle.

2. A particle of mass m moves in the region $x > 0$ under a force having potential

$$V = \lambda(x - a)^2/x^3,$$

where λ and a are positive constants. Sketch the graph of $V(x)$ against x. The particle passes the point $x = \frac{3}{2}a$, travelling in the negative x direction with speed v_0. Discuss all possible subsequent motions of the particle, stating the conditions on v_0 for each different type of motion.

3. A particle of mass m moves on a straight line under the action of a force with potential

$$V = \frac{kx}{x^2 + a^2},$$

where a and k are positive constants. Find the position of stable equilibrium. The particle passes the position of stable equilibrium with velocity u. Obtain an expression for the total energy E of the particle. Find conditions on u for

(i) oscillatory motion

(ii) the particle to escape to $x = -\infty$

(iii) the particle to escape to $x = +\infty$.

4. A particle of mass m moves on a straight line under the action of a force whose potential energy is given by

$$V = ax^2 - bx^3,$$

where a and b are positive constants.

(a) Find the force

(b) Sketch the graph of V against x

(c) The particle passes the origin $x = 0$ with velocity v_0. Show that, if $v_0^2 < 8a^3/27mb^2$, the particle will remain confined to a finite region containing the origin.

5. A particle moving on a straight line is subject to a force

$$F = -k(x - \frac{a^4}{x^3}), \text{ with } k > 0.$$

Find the potential $V(x)$ and sketch its graph. Discuss the nature of the motion of the particle.

2.6 The Gravitational Acceleration

At the end of section 2.1 it was stated that a constant acceleration g provides the classical mathematical model for the motion of a particle moving close to the earth's surface under the action of the earth's gravitational pull and at the end of section 2.2 the corresponding gravitational force acting on a particle of mass m was defined to be the weight of the particle. This gravitational force is of magnitude mg and acts vertically downwards. This model of gravitation is based on experiments conducted by the Italian astronomer and theoretical philosopher Galileo Galilei (1564–1642). Legend has it that Galileo dropped bodies of different masses from the top of the Leaning Tower of Pisa and observed their subsequent motions to be identical. It follows from this that their accelerations must be independent of their masses, so that the constant acceleration g has the same value for all particles. The full significance of this particular experiment will become apparent later, suffice it to say here that the experiment has a central role to play in mathematical models of gravitation and that it has been refined by experimental physicists throughout this century.

The name Galileo will appear elsewhere in this book. In 1585 Galileo faced a difficulty common to many students, namely lack of funding, and withdrew from the University of Pisa before taking his degree. He was responsible for improving and developing the design of the telescope, an instrument which had only recently been introduced. His telescopes were in great demand throughout Europe and he himself used them to great effect in astronomical research. He was appointed in 1610 to a professorship for life. This appointment was made by the Venetian Senate in recognition of his considerable contribution to research and it provided Galileo with financial security. Astronomical research inevitably led him into conflict with the Church. In 1616 the Holy Office decreed the proposition that the earth orbits about the sun to be heretical and Galileo was ordered not to hold, teach or defend such a viewpoint. This he promised to do but in 1632 he published a book which was clearly in contradiction of the edict of 1616 and which broke the promise Galileo had given at that time. Sale of the book was banned and Galileo was brought before the Inquisition. He was forced to spend the last eight years of his life in strict seclusion although nothing, including blindness, diminished his scientific work.

In this section we shall see how the constant gravitational acceleration g can be derived using the concept of linear approximation as introduced in section 1.6. To fix ideas consider a particle P of mass m moving on a vertical line under the earth's gravitational attraction. Let x be the height of the particle above the earth's surface. Assuming that the particle cannot penetrate the earth's surface this height is just the displacement of the particle relative to an origin on the earth's surface, the orientation of the vertical line on which the particle moves being as indicated in Fig 2.14. If the displacement had been defined relative to an origin O not lying on the earth's surface then a negative displacement would indicate that the particle is below O, the depth below being equal to $|x|$.

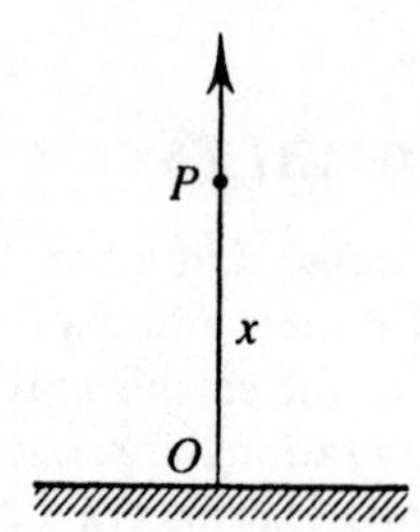

Fig 2.14 Vertical motion of a particle *P*.

The gravitational force of attraction acting on a particle depends on its height above the earth's surface and is therefore a function of its displacement x. The energy equation of the particle P therefore takes the form

$$\tfrac{1}{2}m\dot{x}^2 + V(x) = E.$$

You will recall that the potential energy of any particle is only defined up to an arbitrary additive constant. It is conventional to take the gravitational potential $V(x)$ to be zero at ground level, that is when $x = 0$. If follows that $V(x)$ can be approximated by a linear function, but following the discussion in section 1.7 you may remember that it is important first to introduce a dimensionless variable $\bar{x}$.

This is done by defining

$$\bar{x} = \frac{x}{R},$$

where R is the radius of the earth. For small values of $\bar{x}$ the potential energy can be approximated as a linear function

$$V = p\bar{x} = \frac{px}{R}.$$

Notice that $\bar{x} \ll 1$ implies that $x \ll R$ so that we are considering a particle moving close to the earth's surface. Substituting this linearized potential energy into the energy equation gives

$$\tfrac{1}{2}m\dot{x}^2 + \frac{px}{R} = E.$$

Differentiating this equation with respect to time, using the chain rule to differentiate $\dot{x}^2$, gives

$$\tfrac{1}{2}m2\dot{x}\frac{d\dot{x}}{dt} + \frac{p\dot{x}}{R} = 0,$$

or

$$m\dot{x}\ddot{x} + \frac{p\dot{x}}{R} = 0.$$

Dividing this equation by the common factor $\dot{x}$ and rearranging gives the following expression for the acceleration of the particle, namely

$$\ddot{x} = -\frac{p}{mR}.$$

This acceleration is constant for each particle P although the value of the constant seems to depend on the mass m of the particle. The significance of Galileo's experiment was that this constant acceleration is, in fact, independent of the mass. An intuitive argument can be given to show that this must be the case. The gravitational potential energy V is actually a function of the mass m of the particle as well as of the displacement x. This means that the factor p appearing in the above linear approximation is itself a function of the mass m. Introducing a dimensionless variable

$$\bar{m} = \frac{m}{M},$$

where M is the mass of the earth, it follows that if it is appropriate to make a linear approximation of the function p when $\bar{m} \ll 1$ then

$$p = k\bar{m} = \frac{km}{M},$$

where k is a constant. Substituting this into the previous expression for the acceleration $\ddot{x}$ gives

$$\ddot{x} = -\frac{km}{mRM} = -\frac{k}{RM}.$$

Notice that the mass m has cancelled showing that the gravitational acceleration is indeed independent of the mass.

From the above it follows that the magnitude g of the gravitational acceleration of a particle moving close to the earth's surface is given by

$$g = \frac{p}{mR}$$

and that the gravitational potential energy of the particle is given by

$$V = \frac{px}{R}.$$

From these it follows that

$$V = mgx$$

so that the gravitational potential energy of a particle is equal to its weight mg multiplied by its height.

Of course if we start with the assumption that the weight of the particle is equal to mg then the gravitational force, which acts downwards, is equal to $-mg$ and the above expression for the potential is easily obtained by integrating

$$\frac{dV}{dx} = mg.$$

TUTORIAL PROBLEM 2.6

Discuss whether it is appropriate to make a linear approximation of the function $p(m)$.

TUTORIAL PROBLEM 2.7

The gravitational potential energy increases with height. What does this imply about the danger from falling objects?

Example 10

One end O of a light elastic string of natural length l_0 and modulus of elasticity λ is attached to a fixed point. The string hangs at rest under the weight of a particle P of mass m attached to its other end. Find the extension of the string.

SOLUTION

The total downward force acting on the particle P is equal to $mg - T$. Since P is at rest this resultant force must be zero. Hence

$$T = mg.$$

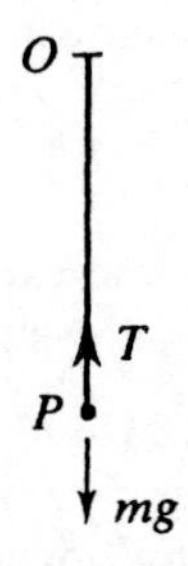

Fig 2.15 Elastic string with particle.

Substituting Hooke's law gives

$$\frac{\lambda x}{l_0} = mg$$

so that the extension of the string is equal to mgl_0/λ.

Example 11

Two light elastic strings AB and CD each of natural length l_0 are joined together at the ends B and C. The end A of the composite string is attached to a fixed point and the composite string hangs at rest under the weight of a particle of mass m attached to the end D. Find the extensions of the two strings, their moduli of elasticity being λ and $k\lambda$.

SOLUTION

The forces acting on the joined ends, and on the particle are shown in Fig 2.16. Since the system is at rest the resultant forces at B and D must both be zero. Hence

$$T = mg$$
$$\text{and} \quad T' = T.$$

If the extensions are x and x' then

$$\frac{\lambda x}{l_0} = mg \quad \text{and} \quad \frac{k\lambda x'}{l_0} = \frac{\lambda x}{l_0}.$$

Fig 2.16 Composite elastic string with particle.

Hence

$$x = \frac{mgl_0}{\lambda} \quad \text{and} \quad x' = \frac{mgl_0}{k\lambda}.$$

You should compare the first of these expressions with the result obtained in Example 10. Why is it the same? Could we have deduced the value of x' from the result obtained in Example 10?

Neglecting air resistance, the only force acting on a particle P of mass m moving under the action of the earth's gravitational attraction alone is its weight. If the motion is vertical then the equation of motion of P is

$$m\ddot{x} = -mg,$$

where the vertical line of motion is orientated upwards and x is the displacement of P relative to a fixed origin O. Cancelling m gives

$$\ddot{x} = -g$$

an equation which we could have written down immediately by considering the kinematics of the situation. Repeated integration of this equation yields the following expressions for the velocity and displacement of the particle relative to O at time t,

$$\dot{x} = -gt + c_1$$

$$\text{and} \quad x = -\tfrac{1}{2}gt^2 + c_1t + c_2.$$

Here c_1 and c_2 are constants of integration. They can be interpreted as the initial velocity and displacement of the particle at time $t = 0$, respectively. The location of the fixed origin O can always be chosen to simplify the problem being solved. For example if the particle is projected vertically upwards then the origin can be chosen to be at the point of projection. Similarly if the particle is dropped, i.e. released from rest, the origin can be chosen to be at the point from which the particle is dropped.

Example 12

A particle is projected vertically upwards from a height h with speed u. Find the maximum height of the particle and the time taken for the particle to fall back to ground level.

SOLUTION

Take the origin at ground level and suppose that the particle is projected at time $t = 0$. Then initially $x = h$ and $\dot{x} = u$. Hence

$$\dot{x} = -gt + u$$

$$\text{and} \quad x = -\tfrac{1}{2}gt^2 + ut + h.$$

The maximum height occurs when $\dot{x} = 0$, i.e. when $t = u/g$. Substituting this into the equation for the displacement x gives the maximum height, namely

$$\frac{u^2}{2g} + h.$$

The particle falls back to the ground when $x = 0$, i.e. when

$$-\tfrac{1}{2}gt^2 + ut + h = 0.$$

This quadratic has two solutions, namely

$$t = \frac{-u + \sqrt{u^2 + 2gh}}{-g} = \frac{u - \sqrt{u^2 + 2gh}}{g}$$

$$\text{and} \quad t = \frac{-u - \sqrt{u^2 + 2gh}}{-g} = \frac{u + \sqrt{u^2 + 2gh}}{g}.$$

The particle cannot fall back to the ground before it has been projected and therefore the required time is given by the second, positive solution. Does the first, negative solution have an interpretation?

EXERCISES ON 2.6

1. A particle is projected vertically upwards from the ground at time $t = 0$ and reaches a height h at time $t = \tau$. Show that the greatest height of the particle is

$$(g\tau^2 + 2h)^2/8g\tau^2.$$

Find also the speed at which the particle is projected.

2. A particle of mass $0.5k$ is projected vertically from the earth's surface with speed $5.0\ \text{ms}^{-1}$. Taking g to be equal to 9.8ms^{-2} obtain a value for the total energy E of the particle. Using the energy equation

$$\tfrac{1}{2}m\dot{x}^2 + mgx = E$$

find the maximum height of the particle and also the speed of the particle when it has fallen half way back to earth.

3. A particle is projected vertically upwards and reaches a maximum height H at time $t = T$. Show that the height of the particle at any time t is given by

$$x = H - \tfrac{1}{2}g(t - T)^2.$$

4. A particle is released from rest and at a time $t = \tau$ later it has fallen a distance d. How much further will the particle have fallen at a time $t = 2\tau$ after release?

5. A particle is projected vertically upwards from a fixed point O with speed u. A short time τ later a second particle is similarly projected from O. Prove that the particles will collide at a time t after the first particle is projected, given by

$$t = \tfrac{1}{2}\tau + \frac{u}{g}.$$

Show that the speeds of the two particles immediately before impact are equal.

2.7 Motion Under a Velocity Dependent Force

Consider a particle of mass m moving on a straight line under the action of a force F which depends only on the velocity $\dot{x}$ of the particle relative to a given fixed origin. The equation of motion becomes

$$m\ddot{x} = F(\dot{x}).$$

It is useful to put $\dot{x} = v$ and to rewrite the equation of motion as

$$m\frac{dv}{dt} = F(v).$$

If $F(v) \neq 0$ this equation can be integrated with respect to the time t to give

$$t = \int \frac{m}{F(v)} dv$$

$$\text{or} \quad t = f(v) + c_1,$$

where $f(v)$ is a particular value of the integral and c_1 is a constant of integration. This equation gives t as a function of v. In principle it can be inverted to give v as a function of t, say $v = G(t)$, where the function G will involve the constant c_1. Thus

$$\frac{dx}{dt} = G(t)$$

and integrating gives

$$x = \int G(t) dt$$

$$\text{or} \quad x = g(t) + c_2$$

where $g(t)$ is a particular value of the integral and c_2 is a second constant of integration. This solution of the equation of motion involves two arbitrary constants and is the general solution; the displacement $x(t)$ describes all possible motions of the particle consistent with the given force $F(\dot{x})$. Alternatively the equation of motion can be rewritten in a form which involves displacement and velocity rather than time and velocity. This is done by using the chain rule

$$\frac{dv}{dt} = \frac{dx}{dt}\frac{dv}{dx} = v\frac{dv}{dx}$$

to rewrite the equation of motion as

$$mv\frac{dv}{dx} = F(v).$$

If $F(v) \neq 0$ this equation can be integrated with respect to the displacement x to give

$$x = \int \frac{mv}{F(v)} dv$$

$$\text{or} \quad x = f'(v) + c_1',$$

where $f'(v)$ is a particular value of the integral and c_1' is a constant of integration. The general solution to the equation of motion can now be deduced by combining the two first integrals

$$t = f(v) + c_1$$
$$\text{and} \quad x = f'(v) + c_1'.$$

One method is to find v as a function of t from the first equation and to substitute the function into the right hand side of the second equation. In terms of the notation already introduced this gives

$$x = f'(G(t)) + c_1'.$$

A second method is to consider the two equations as giving the general solution of the equation of motion in parametric form. The graph of x against t can be drawn by plotting corresponding values of x and t found from the two equations for various values of v.

In applications the force $F(\dot{x})$ will usually be the drag force acting on a particle due to its motion through a resisting medium. This sentence requires some explanation because a particle, having zero size, surely cannot experience a drag force? The sentence presupposes that the particle is modelling a body of finite size which has to be taken into account, together with its shape etc, when modelling the drag. We saw in Section 1.7 that drag forces have two common properties. Their magnitude is a function of the speed of the body or, in the present context, the speed of the particle relative to the resisting medium and they are directed in the opposite sense to the direction of motion of the particle. If the resisting medium is at rest then a fixed origin O will be fixed relative to the medium, so that the speed of the particle relative to the medium will be $|\dot{x}|$, where x is the displacement of the particle relative to O. The direction of the drag force will be determined by the sign of $-\dot{x}$. It follows that the drag force is indeed a function of the velocity $\dot{x}$.

Consider a particle P which is released from rest at time $t = 0$ in a resisting medium and subsequently falls under the action of the drag force and its own weight. The magnitude and direction of each of these forces is shown in Fig 2.17. Let x be the displacement of the particle relative to the point of release O, the line of motion

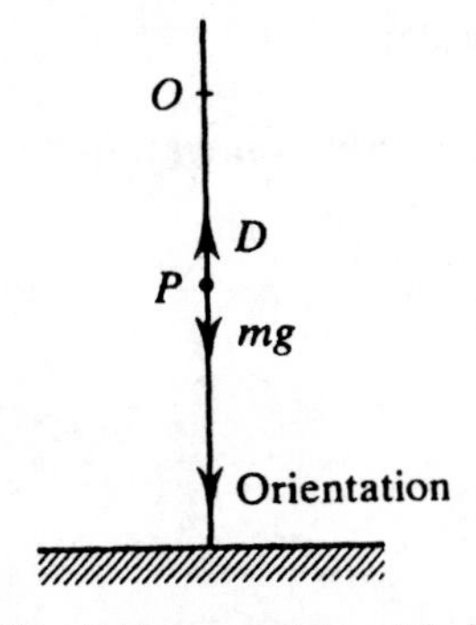

Fig 2.17 Falling particle with drag.

being orientated downwards. In the absence of a drag force the equation of motion would be

$$m\ddot{x} = mg.$$

Note that an orientation different to that in Section 2.6 is being used here. Cancelling the mass m and integrating gives

$$\dot{x} = gt + \text{constant}.$$

Since $\dot{x} = 0$ when $t = 0$ the constant of integration is zero so that

$$\dot{x} = gt.$$

It follows that the velocity of the particle increases indefinitely with time. When the particle hits the ground its velocity is reduced, almost instantaneously, to zero. We saw in Section 1.2 that such a change in velocity is associated with the action of an impulsive force. If the particle is released from a great height, this impulsive force will be large and will cause damage to the particle. It is very interesting to see what happens when the drag force is included in the equation of motion to give

$$m\ddot{x} = mg - D.$$

Assuming that the drag can be modelled as a Stokes' drag it follows from Section 1.8 that $D = 3\pi\eta dv$. The speed v is equal to $|\dot{x}|$. Here the particle is falling in the positive x direction so that $\dot{x} > 0$ and $v = \dot{x}$. The equation of motion can therefore be written as

$$m\ddot{x} = mg - 3\pi\eta d\dot{x}.$$

It is convenient to put $3\pi\eta d = mk$, where k is a positive constant. The equation of motion then simplifies, after cancelling the common factor m, to the form

$$\ddot{x} = g - k\dot{x}$$

so that

$$\frac{d\dot{x}}{dt} = k\left(\frac{g}{k} - \dot{x}\right).$$

Initially $\dot{x} = 0$ so that $g/k > \dot{x}$ during the first period of the particle's fall. The above equation can therefore be integrated to give

$$\log\left(\frac{g}{k} - \dot{x}\right) = -kt + \text{constant}.$$

Since $\dot{x} = 0$ when $t = 0$ the constant of integration is equal to $\log\frac{g}{k}$ so that

$$\log\left(\frac{g}{k} - \dot{x}\right) = -kt + \log\left(\frac{g}{k}\right)$$

$$\text{or} \quad \log\left(\frac{\frac{g}{k} - \dot{x}}{\frac{g}{k}}\right) = -kt.$$

Inverting this equation yields

$$\dot{x} = \frac{g}{k}(1 - e^{-kt}).$$

It follows from this that the speed of the particle is increasing but not now indefinitely. As $t \to \infty$, $\dot{x} \to g/k$ and it follows that the speed of the particle can never exceed the value g/k. This value is called the **limiting speed** of the particle.

For large values of k the limiting speed might be quite small and then it is possible for the particle to hit the ground without being damaged, even when released from a great height. The existence of a limiting speed is the principle behind the working of a parachute and also explains why dogs and cats are often found alive after having fallen down disused mineshafts etc. However in Section 1.8 we saw that Stokes' drag is often an inappropriate model for the drag on a body falling through air and this prompts the question as to whether a limiting speed exists for drags other than Stokes' drag? For a general drag $D(|\dot{x}|)$ the equation of motion is

$$m\ddot{x} = mg - D(|\dot{x}|).$$

As the speed tends to its limiting constant value the acceleration $\ddot{x}$ tends to zero. The limiting speed can therefore be found by setting $\ddot{x} = 0$; you may check this for Stokes' drag. Hence the limiting speed will be the smallest solution $|\dot{x}|$ to the algebraic equation

$$D(|\dot{x}|) = mg,$$

if such a solution exists. If this solution is small enough then the resulting limiting speed will again enable a parachutist to land safely!

TUTORIAL PROBLEM 2.8

The equation $\ddot{x} = g - k\dot{x}$ was obtained for a particle falling under gravity in a medium whose resistance is modelled as a Stoke's drag. Discuss whether the equation needs modification if the particle is projected vertically upwards, rather than released from rest, no change being made to the orientation of the line of motion.

Example 13

The equation of motion of a particle of mass m which is released from rest at the point $x = 0$ in a resisting medium is

$$m\ddot{x} = mg - mk\dot{x},$$

where k is a positive constant. Show that the limiting speed of the particle is given by $v_0 = g/k$. Prove that during the motion

$$x = \frac{1}{k}\left\{v_0 \log\left(\frac{v_0}{v_0 - v}\right) - v\right\},$$

where $v = \dot{x}$ is the velocity of the particle.

SOLUTION

The limiting speed is given by putting $\ddot{x} = 0$. Then $\dot{x} = g/k$ and so the limiting speed $v_0 = g/k$. Putting $\dot{x} = v$ and eliminating m and g, the equation of motion becomes

$$\frac{dv}{dt} = k(v_0 - v).$$

The required result relates x and v and so this equation has to be rewritten in a form in which the time t does not appear explicitly. This is done by using the chain rule

$$\frac{dv}{dt} = \frac{dx}{dt}\frac{dv}{dx} = v\frac{dv}{dx}$$

to rewrite the equation as

$$v\frac{dv}{dx} = k(v_0 - v).$$

From this

$$x = \int \frac{v\,dv}{k(v_0 - v)}$$
$$= \frac{1}{k}\int \left[-1 + \frac{v_0}{v_0 - v}\right] dv.$$

Remembering that $v < v_0$ this can be integrated to give

$$x = \frac{1}{k}[-v - v_0 \log(v_0 - v)] + \text{constant}.$$

Substituting the initial conditions $x = 0$, $v = 0$ into this equation gives the value of the constant, namely

$$\text{constant} = \frac{v_0}{k}\log v_0.$$

Hence

$$x = \frac{1}{k}[-v - v_0 \log(v_0 - v) + v_o \log v_0]$$
$$= \frac{1}{k}[v_0 \log(\frac{v_0}{v_0 - v}) - v],$$

as required.

EXERCISES ON 2.7

1. Write down the equation of motion for a particle of mass m falling vertically in a medium whose resistance to motion is proportional to the speed of the particle (take the constant of proportionality to be mk and use x to denote the height of the particle). The particle is released from rest at time $t = 0$ from a height h. Prove that at time t

$$x = -\frac{g}{k^2}e^{-kt} - \frac{g}{k}t + \frac{g}{k^2} + h.$$

2. A ball bearing of mass 10gm and radius 0.2cm is dropped into a beaker of castor oil. Assuming that the Reynold's number of the motion is small compared to

unity use the data in Section 1.8 to find the limiting speed of the ball bearing. Will this limiting speed agree with experiment?

3. A particle is projected vertically downwards in a medium whose resistance can be modelled as a Stokes' drag with a velocity greater than the limiting speed (the orientation has been chosen so that the velocity is positive). Discuss the subsequent motion of the particle.

Summary

- the **displacement** of a particle P relative to an origin or observer O is a directed number which specifies both the **distance** of P from O and the **orientation** of P relative to O
- the **velocity** of P relative to O is the derivative of the displacement
- the **acceleration** of P relative to O is the derivative of the velocity
- the modulus of velocity is **speed** and the sign of velocity indicates the direction of motion
- the following addition laws hold

$$\text{disp of } P \text{ rel to } O' = \text{disp of } P \text{ rel to } O + \text{disp of } O \text{ rel to } O'$$
$$\text{vel of } P \text{ rel to } O' = \text{vel of } P \text{ rel to } O + \text{vel of } O \text{ rel to } O'$$
$$\text{acc of } P \text{ rel to } O' = \text{acc of } P \text{ rel to } O + \text{acc of } O \text{ rel to } O'$$

- a particle is **at rest** relative to O if and only if its velocity is zero
- a particle is **stationary** or **instantaneously at rest** relative to O at a given instant of time if and only if its velocity is zero at that instant
- a point at which a particle is stationary is called a **stationary point**
- the **equation of motion** $m\ddot{x} = F(t, x, \dot{x})$ expresses Newton's second law of motion for a particle of mass m moving on a straight line under the action of a force F
- if more than one force is acting then F is the **resultant** force obtained by adding all the individual forces
- the general solution of the equation of motion describes all **possible** motions consistent with the given force F. The **actual** motion of a given particle can only be determined if the equation of motion is augmented by some **initial conditions**
- the displacement and velocity at a given instant of time constitute a set of initial conditions; different sets exists
- the **potential** of a force $F(x)$ or **potential energy** of a particle moving under the action of the force is defined by the equation $F(x) = -dV/dx$
- the **energy equation** $\frac{1}{2}m\dot{x}^2 + V(x) = E$ expresses the **conservation of the total energy**, $\frac{1}{2}m\dot{x}^2$ being the **kinetic** energy
- $\int_{x_0}^{x} F(x)dx$ is the **work done** by the force in moving from x_0 to x

- **power** is the rate of change of work done with respect to time
- the potential energy stored in an elastic string is equal to $\lambda x^2/2l_0$, it being assumed that Hooke's law is valid
- the **weight** of a particle of mass m is equal to mg where g is the constant gravitational acceleration
- the **gravitational potential energy** of a particle of mass m moving close to the earth's surface is equal, in the linear approximation, to its weight times its height above a chosen point
- **turning points** are points at which the motion of a particle changes direction
- positions of equilibrium are points at which the force $F(x)$ is zero, so that the potential $V(x)$ is stationary
- positions of equilibrium are **stable/unstable** according as to whether the potential is a minimum/maximum
- the **limiting speed** of a particle falling under a drag force is obtained by putting the acceleration equal to zero; in the case of Stokes' drag the limiting speed is g/k, where $k = 3\pi\eta d/m$.

FURTHER EXERCISES

1. A lift accelerates uniformly from rest with acceleration a until it achieves a speed v. It continues with this speed v and then decelerates to rest on reaching the next floor, at the same rate as it accelerated. If the distance between floors is d, show that the time during which it is travelling at constant speed is given by

$$\left(T^2 - \frac{4d}{a}\right)^{1/2},$$

where T is the total time of travel between floors.

2. A friend wrote to Albert Einstein with the following "brainteaser":

> "An old broken down car has to travel a two mile route uphill and down. Because it is so old, it can take the first mile – the ascent – no faster than at an average speed of 15 m.p.h. Question: How fast does it have to cover the second mile – on the descent it can go faster, of course – in order to achieve an average speed (for the whole distance) of 30 m.p.h?"

In his reply Einstein admitted that he had been fooled into giving the wrong answer. Can you do better?

3. A particle of mass m moves in the region $x > 0$ under a force having potential

$$V(x) = \frac{a}{x^3} - \frac{b}{x^2},$$

where a and b are positive constants. Sketch the graph of $V(x)$ against x. Hence prove that if the particle passes the point $x = 3a/2b$ with a non zero speed v_0

such that $v_0^2 < 8b^3/27ma^2$ then the motion will be oscillatory. What happens if $v_0^2 \gg 8b^3/27ma^2$?

4. In Newtonian cosmology the radius R of the universe is a function of time satisfying the equation

$$\dot{R}^2 = \frac{k^2}{R} + \frac{\lambda R^2}{3} + E,$$

where k, λ and E are constants. Sketch the graphs of the function

$$V(R) = -\frac{k^2}{R} - \frac{\lambda R^2}{3}$$

for $R > 0$ in the cases when $\lambda > 0$, $\lambda = 0$ and $\lambda < 0$. Use these graphs to show that if the universe is expanding from a small radius then it will continue to expand indefinitely if and only if either

$$\lambda > 0 \quad \text{and} \quad 4E^3 > -9\lambda k^4, \quad \text{or} \quad \lambda = 0 \quad \text{and} \quad E \geq 0.$$

5. A weighing machine is constructed by supporting a platform of negligible mass on a spring of natural length a and modulus λ, the other end of the spring being fixed. The weight mg of a body is measured by x, the contraction of the spring when the body is placed on the platform. Show that if the body is hanging at the end of an elastic string of natural length b and modulus λ' when it is placed on the platform and if the other end of the elastic string is attached to a fixed point a distance $a + b$ above the fixed end of the spring then the apparent weight of the body will be reduced by

$$\frac{a\lambda'}{a\lambda' + b\lambda} \times 100\%.$$

6. A light elastic string AB consists of two portions AC and CB each of natural length a, the moduli of elasticity being λ_1 and λ_2, respectively. The string hangs vertically from the fixed end A, the end B carrying a particle of mass m. Show that the equilibrium position of the particle is as if the elastic string AB carrying the particle was of natural length $2a$ and modulus of elasticity λ given by

$$\frac{2}{\lambda} = \frac{1}{\lambda_1} + \frac{1}{\lambda_2}.$$

A second particle of mass m is now attached at the point C. Show that the new equilibrium position of the particle at B is as if the elastic string AB, still carrying both particles, was of natural length $2a$ and modulus of elasticity λ given by

$$\frac{3}{\lambda} = \frac{2}{\lambda_1} + \frac{1}{\lambda_2}.$$

7. A particle is projected vertically, but a screen obscures the actual projection. Can either the time, height, or speed of projection be deduced from subsequent observations?

8. By writing $\ddot{x} = \dot{x}\frac{d\dot{x}}{dx}$ show that the equation

$$\ddot{x} = -g - k\dot{x}$$

for the motion of a particle falling under gravity and the action of a Stokes' drag can be integrated with respect to the height x to yield

$$\dot{x} - \frac{g}{k}\log_e\left(\frac{k\dot{x}}{g} + 1\right) = -k(x - h),$$

the particle being released from rest at a height $x = h$. Hence prove that the particle attains $p\%$ of its limiting speed after falling a distance

$$d = -\frac{g}{k^2}\left\{\frac{p}{100} + \log_e\left(1 - \frac{p}{100}\right)\right\}.$$

Viscous oil of density $\rho = 1.0 \times 10^3\text{km}^{-3}$ and viscosity $\eta \approx 12k\text{m}^{-1}\text{s}^{-1}$ fills a beaker 1.0×10^{-1} m deep. A ball bearing of mass $m = 1.0 \times 10^{-2}k$ and radius $a = 2.5 \times 10^{-3}$ m is released from rest at the free surface of the oil. Assuming that the resistance of the oil can be modelled as a Stokes' drag show that the ball bearing attains 99% of its limiting speed before falling through 1.1×10^{-2} m of the oil.

9. The drag D on a sky diver of mass m is modelled by the expression

$$D = mcv^2$$

where c is a constant and v is the speed of the sky diver. Find an expression for the limiting speed v_0 and show that after falling for a time t the speed of the sky diver is given by

$$v = v_0\frac{e^{2gt/v_0} - 1}{e^{2gt/v_0} + 1}.$$

A typical limiting speed is 54ms^{-1}. Show that the sky diver attains 50% and 90% of the limiting speed after 3.1s and 8.2s of fall respectively.

3 • Simple Harmonic Motion

Discussion of the motion of a particle close to a position of stable equilibrium leads to the simple harmonic motion equation. The general solution of this equation and of other equations appearing in this Chapter is obtained by ad hoc methods. These equations are all second order linear differential equations and readers who have already studied differential equations are urged to rederive the general solutions using the standard methods which they will have learnt. The general solution of the simple harmonic motion equation is shown to represent an oscillation and the effect on this oscillation of the presence of a resisting medium is investigated under the assumption that the resistance is modelled as a Stokes' drag. It is found that the oscillator then always tends to its equilibrium position as the time tends to infinity. This is known as damping and is modelled by the damped simple harmonic motion equation. A sinusoidal forcing term is introduced into the equation and the properties of the resulting forced damped oscillations are discussed.

3.1 Motion Close to a Position of Stable Equilibrium

Positions of stable and unstable equilibrium were introduced in Section 2.5 as being points at which the potential energy $V(x)$ of a particle is a minimum or maximum, respectively. At such points dV/dx is necessarily zero and so the force $F(x) = -dV/dx$ acting on the particle is also zero. A particle placed at rest at a position of equilibrium will therefore remain at rest; the particle is then said to be in stable/unstable equilibrium.

If a particle in stable equilibrium is disturbed slightly either by displacing it to a nearby position or by applying a small impulsive force to start it moving, then its subsequent motion will consist of an oscillation about the position of equilibrium. The particle will not move far from its position of stable equilibrium. By contrast a particle in unstable equilibrium will, if disturbed slightly, move right away from its position of equilibrium.

TUTORIAL PROBLEM 3.1

The above discussion was first given in Section 2.5 and based on the analysis of the graph of the potential energy close to each position of equilibrium, that is close to each stationary point of the function $V(x)$. Discuss the motion of a particle disturbed from rest at a stationary point of $V(x)$ which is a point of inflexion.

In theory you should be able to balance a flat headed screw, or nail, upright on its point. The screw would be in a position of equilibrium. In practice you will inevitably fail to do so because this position of equilibrium is unstable and so if you release the screw at even the smallest of angles to the vertical it will move right away from the position of equilibrium and fall down. The task is far easier if you try to balance the screw on its flat end. This corresponds to a position of stable equilibrium. You will probably notice that when you release the screw it wobbles slightly but then settles down in an upright position. Of course if you are really clumsy the screw will again fall down but that is because you are not releasing it close enough to its position of equilibrium. Most of the objects about us are in equilibrium, usually in stable equilibrium. Two particular examples which can usefully be adapted to both thought and actual experimentation are bodies suspended at rest at the end of inextensible strings – light pulls, chains on old toilet cisterns, pulls on roller blinds, plumblines, children's swings – and bodies suspended at rest at the end of elastic strings or springs – simple spring scales used for weighing, exhausted bungee divers; perhaps you can add to these lists? In the first examples the body can be disturbed from its equilibrium position by moving it so that the string, although taut, is no longer vertical. When released the body will swing about its position of equilibrium. Of course this motion is not straight line motion. This suggests that the concepts of equilibrium and stability will be of importance in situations other than straight line motion. In the second examples the body can be disturbed from its equilibrium position by displacing it vertically. When released it will oscillate on a vertical line about its position of equilibrium. Neither the swinging nor the oscillation of the bodies discussed above are in any way random. Both are regular and are examples of a special motion, called simple harmonic, which characterizes the motion of any particle close to a position of stable equilibrium.

TUTORIAL PROBLEM 3.2

(i) Imagine a small body suspended by a spring from a fixed point. The body is pulled down and released at time $t = 0$ from the point $x = a > 0$. Here the displacement x is defined relative to the equilibrium position of the body and the vertical line of motion is orientated downwards. Plot the maximum, minimum and zero values of the displacement against time for several oscillations.

(ii) Think about the velocity of the body now. Plot the velocity against time for the same times as in part (i). It might help if you think about when the velocity is a maximum, a minimum and zero.

(iii) Now consider the acceleration. When is it a maximum, a minimum and zero? Plot the acceleration against time for the same times as in parts (i) and (ii). Imagine curves drawn through the plotted points. Recalling that velocity is the derivative of displacement and that acceleration is the derivative of velocity, what functions of time t might the displacement, velocity and acceleration be?

Example 1

A particle of mass m is attached to one end of a light spring of modulus λ and natural length l_0, the other end of which is attached to a fixed point. Write down the equation of motion for vertical oscillations of the particle using (i) the position of equilibrium and (ii) the fixed end of the spring as the origin, the orientation of the vertical line of motion being downwards in both cases.

SOLUTION

(i) Let x be the displacement of the particle relative to the position of equilibrium O. The extension of the spring is $d + x - l_0$, where d is the displacement of O relative to the point of suspension. The tension in the spring is given by

$$T = \frac{\lambda}{l_0}(d + x - l_0).$$

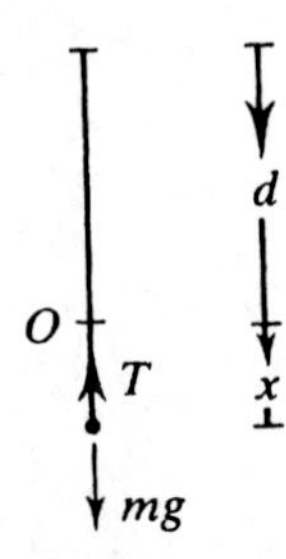

Fig 3.1 Oscillation relative to the equilibrium position.

The forces acting on the particle are directed as in Fig 3.1 and therefore the equation of motion is

$$m\ddot{x} = mg - \frac{\lambda}{l_0}(d + x - l_0).$$

This equation can be written in the form

$$\ddot{x} + \frac{\lambda x}{ml_0} = g - \frac{\lambda}{ml_0}(d - l_0).$$

The right hand side of this equation is a constant. Since $\ddot{x} = 0$ at the position of equilibrium $x = 0$ it follows that this constant must be zero. This leads to an equation from which the displacement d of the position of equilibrium can be found to be

$$d = l_0 + \frac{ml_0 g}{\lambda}$$

and to a simple form of the equation of motion, namely

$$\ddot{x} + \frac{\lambda x}{ml_0} = 0.$$

This derivation clearly demonstrates that it is possible to obtain the simplified form of the equation of motion without first finding the position of equilibrium. An alternative derivation and perhaps a more natural approach, involves finding the position of equilibrium before writing down the equation of motion.

(i)′ Let x_0 be the extension of the spring when the particle is hanging in equilibrium at the point O. In Fig 3.1 d now becomes equal to $l_0 + x_0$. In equilibrium the total force acting on the particle is zero and so, taking into account the directions of the tension and weight it follows that $T = mg$ or, using Hooke's law,

$$\frac{\lambda}{l_0}x_0 = mg.$$

This determines x_0 and therefore the position of the point of equilibrium O. Now take O as the origin and let x be the displacement of the particle relative to O. The extension is $x_0 + x$ so that

$$T = \frac{\lambda}{l_0}(x_0 + x)$$

and the equation of motion is

$$m\ddot{x} = mg - \frac{\lambda}{l_0}(x_0 + x).$$

Eliminating x_0 this becomes

$$m\ddot{x} = -\frac{\lambda}{l_0}x$$

which can be rewritten in the form

$$\ddot{x} + \frac{\lambda}{ml_0}x = 0.$$

(ii) Now let x be the displacement of the particle relative to the point of suspension O. The extension of the spring is given by $x - l_0$ and the equation of motion becomes

$$m\ddot{x} = mg - \frac{\lambda}{l_0}(x - l_0)$$

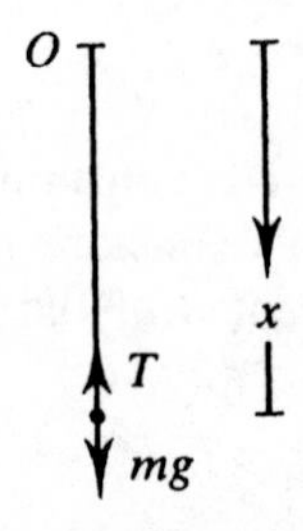

Fig 3.2 **Oscillation relative to the point of suspension.**

which can be written in the form

$$\ddot{x} + \frac{\lambda}{ml_0}x = g + \frac{\lambda}{m}.$$

The right hand side of this equation is again a constant but, unlike the analogous constant in part (i), it is non zero. Putting $\ddot{x} = 0$ the corresponding value of x gives the displacement of the position of equilibrium relative to the point of suspension, the quantity which was denoted by d in part (i). •

Consider a particle of mass m moving close to a given position of equilibrium. Let x be the displacement of the particle relative to this position of equilibrium. The force $F(x)$ is zero at $x = 0$ and for small displacements, $|x| \ll 1$, this force can be approximated as a linear force,

$$F(x) = px$$

where the constant p is the slope of the graph of $F(x)$ at $x = 0$, i.e. $dF/dx|_0$. Here $|_0$ denotes evaluation at $x = 0$. To avoid ambiguity in the inequality $|x| \ll 1$ a dimensionless displacement should be used here, obtained by dividing the displacement by a length characteristic of the motion under consideration, see Section 1.7. Since $F = -dV/dx$ the constant p can also be written as $-d^2V/dx^2|_0$. With this linear approximation, the equation of motion becomes

$$m\ddot{x} = -\frac{d^2V}{dx^2}\bigg|_0 x.$$

At a position of stable equilibrium $V(x)$ is a minimum so that $d^2V/dx^2|_0$ is positive. Defining $w > 0$ by the equation

$$w^2 = \frac{1}{m}\frac{d^2V}{dx^2}\bigg|_0$$

the equation of motion can be rewritten as

$$\ddot{x} + w^2x = 0.$$

Multiplying this by $\dot{x}$ and integrating with respect to time gives

$$\tfrac{1}{2}\dot{x}^2 + \tfrac{1}{2}w^2x^2 = E',$$

where E' is a constant, the total energy per unit mass. The graph of the potential energy per unit mass, $V' = \frac{1}{2}w^2x^2$, is shown in Fig 3.3. Motion can only take place close to $x = 0$ if $E' > 0$ and then that motion is oscillatory. With suitable initial conditions the turning points will be reached before the displacement becomes large enough for the linear approximation to break down.

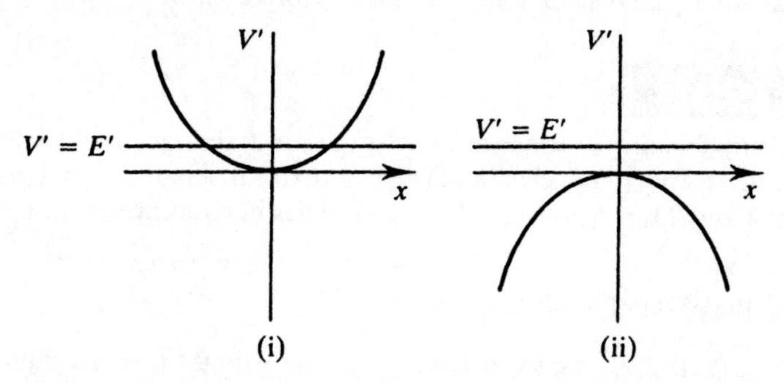

Fig 3.3 Linear approximant of the potential energy per unit mass.

At a position of unstable equilibrium $V(x)$ is a maximum so that $d^2V/dx^2|_0$ is negative. Defining $w > 0$ by the equation

$$w^2 = -\frac{1}{m}\frac{d^2V}{dx^2}\Big|_0$$

the equations of motion and conservation of energy per unit mass become

$$\ddot{x} - w^2x = 0$$

and $\frac{1}{2}\dot{x}^2 - \frac{1}{2}w^2x^2 = E'$,

respectively. The graph of the potential energy per unit mass in this case is shown in Fig 3.3(ii). The motion is no longer oscillatory, instead the particle will escape to $\pm\infty$. Of course the linear approximation is only applicable for small displacements and will break down as the particle moves away from the origin.

Example 2

The equation of motion obtained in Example 1 part (i) is of the form

$$\ddot{x} + w^2x = 0.$$

Find an expression for w^2 by evaluating $\frac{1}{m}\frac{d^2V}{dx^2}\big|_0$.

SOLUTION

The total force acting on the particle is

$$F(x) = mg - \frac{\lambda}{l_0}(d + x - l_0)$$

$$= \text{constant} - \frac{\lambda}{l_0}x.$$

Now $F(x) = -dV/dx$ and therefore

$$\frac{d^2V}{dx^2} = \frac{\lambda}{l_0}.$$

Hence $w^2 = \lambda/ml_0$ as was obtained in Example 1. Notice that d^2V/dx^2 is a constant in this example and that the equation of motion is exact rather than being a linear approximation. This is because a linear approximation has already been used to model the tension in the spring by Hooke's law. •

EXERCISES ON 3.1

1. A particle P of mass m is attached to two points A and B on a smooth horizontal table, with $AB = 5l$, by elastic strings of natural lengths l and moduli of elasticity λ and 2λ, respectively. The particle lies in equilibrium on the straight line AB.

   (i) Find the point of equilibrium O.

   (ii) The particle P is displaced from O on the straight line AB and released. Show that, provided both strings remain taut, the equation of motion of P is of the form

$$\ddot{x} + w^2 x = 0,$$

where $w^2 = 3\lambda/lm$

and x is the displacement of P from O, measured positively in the direction AB.

2 Write down the total force $F(x)$ acting on the particle P in question 1 and hence show that $w^2 = 3\lambda/lm$ by evaluating $\frac{1}{m}\frac{d^2V}{dx^2}\Big|_0$.

3 A particle of mass m is moving on a straight line under the action of a force whose potential energy $V(x)$ is given in SI units by

$$V(x) = 3x^2 - 2x^3.$$

(i) Sketch the graph of $V(x)$ against x.

(ii) Identify the direction of the force $F(x)$ acting on the particle for each value of the displacement x.

(iii) Find the values of the total energy E such that oscillatory motion is possible.

(iv) Show that if the amplitude of the oscillatory motion is small then that motion is determined by the equation

$$\ddot{x} + w^2 x = 0,$$

where $w^2 = 6/m$.

(v) What are the units of the coefficients 3 and 2 appearing in the expression for $V(x)$?

3.2 Simple Harmonic Motion

The equation

$$\ddot{x} + w^2 x = 0$$

is called the **simple harmonic motion equation** and any mechanical system whose motion is determined by this equation is called a **simple harmonic oscillator**. Simple harmonic motion is of great importance because, as we saw in the last section, it is the motion experienced by any particle oscillating close to a position of stable equilibrium. If the displacement x is defined relative to an origin other than the position of equilibrium then a constant will appear on the right hand side of the equation, as in Example 1.

The simple harmonic motion equation is a second order linear homogeneous differential equation. You may already have studied differential equations. If so you will know at least one standard method for solving such equations and you should apply this method to the simple harmonic equation. For those readers who have not yet studied differential equations we will give a non rigorous derivation of the general solution to the equation based on the observation that the two functions $x = \sin wt$ and $x = \cos wt$ each satisfy the equation. This is easily verified by substitution. For example, if $x = \sin wt$ then $\dot{x} = w\cos wt$ and $\ddot{x} = -w^2 \sin wt$ so that $x + w^2 x = 0$. Because the simple harmonic motion equation is linear it follows that a linear combination of these two solutions, with arbitrary constant coefficients,

$$x = c_1 \sin wt + c_2 \cos wt$$

is also a solution (this can be verified directly by substitution). This solution contains two arbitrary constants which can be determined given suitable initial conditions. It is the required general solution.

The general solution of the simple harmonic motion equation can be written in several alternative forms and in particular in the form

$$x = A\sin(wt + \delta),$$

where A and δ are constants with $A \geq 0$ and $0 \leq \delta < 2\pi$. To see this use the trigonometrical identity for the sum of two angles to write the above as

$$x = A\sin wt \cos\delta + A\cos wt \sin\delta.$$

This is in the form of the general solution obtained above, with

$$A\cos\delta = c_1 \quad \text{and} \quad A\sin\delta = c_2.$$

From these equations A and δ can be found in terms of c_1 and c_2. Squaring and adding gives

$$A^2 = c_1^2 + c_2^2.$$

Since A is positive it follows that, provided $A \neq 0$,

$$\cos\delta = c_1/(c_1^2 + c_2^2)^{1/2} \quad \text{and} \quad \sin\delta = c_2/(c_1^2 + c_2^2)^{1/2}$$

giving a unique value for δ lying in the interval $0 \leq \delta < 2\pi$. Note that both equations, or their equivalent are required to obtain δ uniquely.

The motion of a simple harmonic oscillator is best discussed by analysing the solution

$$x = A\sin(wt + \delta).$$

Because the sine function is periodic the motion is indeed oscillatory. In fact

$$A\sin(wt + \delta) = A\sin(wt + \delta + 2\pi) = A\sin(w\{t + 2\pi/\omega\} + \delta)$$

from which it follows that the oscillation is regular, the time taken for one complete oscillation being equal to $2\pi/w$. This is known as the **periodic time** or **period of oscillation**. The inverse of the periodic time, $w/2\pi$, is the number of complete oscillations performed per unit time. This is called the **frequency**, w itself being called the **angular frequency**. The SI unit of frequency is the hertz (Hz), one hertz being a frequency of one oscillation per second ($\text{Hz} = \text{s}^{-1}$). The sine function is bounded, its value lying between ± 1, and so the motion of the simple harmonic oscillator is bounded, its displacement lying between $\pm A$. A is called the **amplitude**, it is the maximum displacement of the oscillator from its position of equilibrium or, alternatively, the maximum distance of the oscillator from its position of equilibrium. The graph of the displacement x against time t is shown in Fig 3.4 for

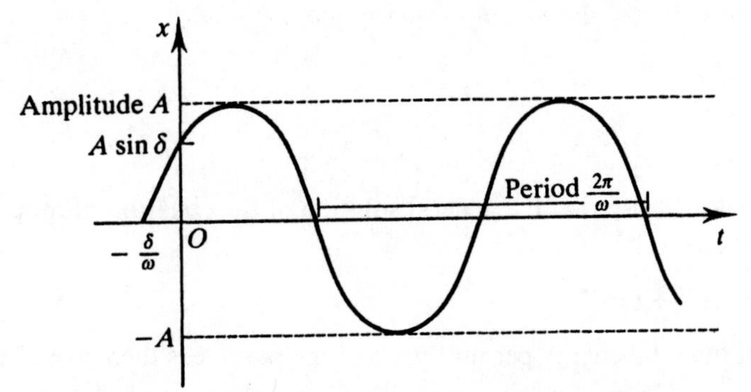

Fig 3.4 Displacement of simple harmonic oscillator.

the case when δ lies in the first quadrant. It has been obtained by shifting the graph of the sine function, which passes through the origin, in the negative time direction. The amount of shift depends on the value of the constant δ. This constant is known as the **phase constant**. The graph of the displacement for the case when δ lies in other than the first quadrant is obtained from Fig 3.4 by moving the Ox axis in the positive time direction.

TUTORIAL PROBLEM 3.3

Consider a particle performing simple harmonic motion on the end of a light spring as in Example 1. Discuss the behaviour of the period of oscillation with (a) the modulus λ (b) the natural length l_0 and (c) the mass m.

The energy equation

$$\tfrac{1}{2}\dot{x}^2 + \tfrac{1}{2}w^2x^2 = E'$$

was obtained in the last section. E' is called the energy of the simple harmonic oscillator. If the oscillator is a particle of mass m then this energy E' is just the total energy per unit mass of the particle. Substituting $x = A\sin(wt + \delta)$ into the energy equation gives

$$\tfrac{1}{2}A^2w^2\cos^2(wt+\delta) + \tfrac{1}{2}w^2A^2\sin^2(wt+\delta) = E',$$

so that

$$E' = \tfrac{1}{2}w^2A^2.$$

The energy of the oscillator is therefore proportional to the squares of the amplitude and frequency of the oscillation.

TUTORIAL PROBLEM 3.4

In Section 3.1 the equation of motion and the energy equation for a particle moving close to a position of unstable equilibrium were found to be

$$\ddot{x} - w^2x = 0$$

and $\frac{1}{2}\dot{x}^2 - \frac{1}{2}w^2x^2 = E'$,

respectively. Verify that the general solution of the equation of motion can be written as

$$x = c_1e^{wt} + c_2e^{-wt}$$

and that the total energy per unit mass of the particle is then given by

$$E' = -2c_1c_2w^2.$$

Discuss the motion of the particle.

Example 3

Consider the particle in Example 1 now suspended by a light elastic string of natural length l_0 and modulus of elasticity λ rather than by a spring. The particle is observed to pass its equilibrium point at time $t = 0$ moving with velocity $v_0 > 0$. It subsequently reaches the highest point of its motion for the first time at $t = t_1$. Prove that if the string remains taut throughout the motion $v_0 < g/w$ and that knowing the mass of the particle and the natural length of the string, the modulus of elasticity can be found from this data.

SOLUTION

So long as the string remains taut the equation of motion of the particle is

$$\ddot{x} + w^2x = 0,$$

where $w^2 = \lambda/ml_0$ and x is the displacement of the particle relative to its position of equilibrium 0. Substituting the initial conditions $x = 0, \dot{x} = v_0$ when $t = 0$ into the general solution $x = A\sin(wt + \delta)$ gives

$$0 = A\sin\delta \quad \text{and} \quad v_0 = wA\cos\delta.$$

From the first equation $\delta = 0$ or π and since $v_0 > 0$ the second equation requires that $\cos\delta > 0$ leaving $\delta = 0$. The second equation now gives $A = v_0/w$. In Example 1 the displacement d of O relative to the point of suspension was found to be given by

$$d = l_0 + \frac{ml_0g}{\lambda}.$$

The highest and lowest points of the oscillation are shown in Fig 3.5. The displacement of the highest point relative to the point of suspension is $d - A$ and if the string remains taut throughout the motion then $d - A$ must be greater than

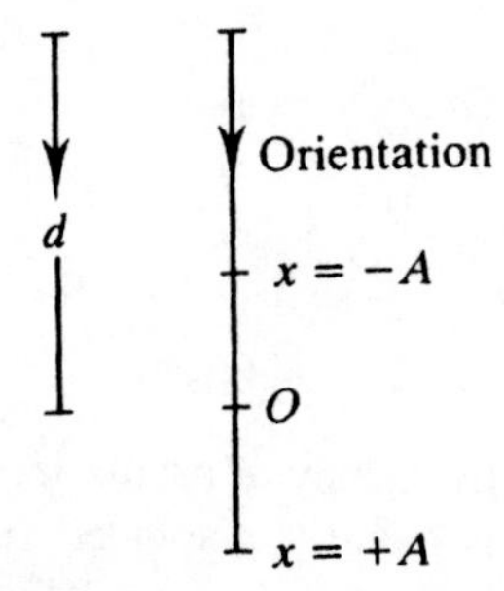

Fig 3.5 Highest and lowest points of oscillation.

or equal to l_0. Substituting for d and A gives

$$l_0 + \frac{ml_0g}{\lambda} - \frac{v_0}{w} \geq l_0$$

$$\text{or} \quad v_0 \leq \frac{ml_0gw}{\lambda}.$$

Using $w^2 = \lambda/ml_0$ this can be written as $v_0 \leq g/w$. The particle passes the highest point when $\sin wt = -1$. The first time this occurs is when $t = t_1 = 3\pi/2w$. Now

$$\lambda = w^2ml_0 = \frac{9\pi^2ml_0}{4t_1^2}$$

and so λ can indeed be determined from the given data. Note that the value of v_0 is not required. •

Example 4

A particle oscillates on a straight line with simple harmonic motion about a point O. The particle passes a point distant 0.10m from O moving towards O with a speed 0.15ms^{-1}. At O the speed of the particle is observed to be 0.26ms^{-1}. Find the period and amplitude of oscillation.

SOLUTION

Since the observed data does not involve time the required values of the period and amplitude can be found from the energy equation using the general result $E' = \frac{1}{2}w^2A^2$. Substituting $|\dot{x}| = 0.15$ when $|x| = 0.10$ and $|\dot{x}| = 0.26$ when $x = 0$ into the equation

$$\frac{1}{2}\dot{x}^2 + \frac{1}{2}w^2x^2 = \frac{1}{2}w^2A^2$$

gives

$$0.15^2 + w^2 0.10^2 = w^2A^2 \text{ and } 0.26^2 = w^2A^2.$$

From these

$$w^2 0.1^2 = 0.15^2 - 0.02^2$$

so that $w = 2.12$.

The period of oscillation is

$$2\pi/w = 2\pi/2.12 = 3.0\text{s}$$

and the amplitude A is

$$0.26/w = 0.26/2.12 = 0.12\text{m}.$$

EXERCISES ON 3.2

1. Obtain an expression for the frequency of oscillation of the particle in question 3 of Exercises 2.5 when it is slightly displaced from its position of stable equilibrium.

2. A particle is describing simple harmonic motion with period 0.7s, moving on a straight line. At $t = 0$, the particle is observed to be a distance 0.08m from the centre of oscillation moving towards it with a speed 0.9ms^{-1}. Find the amplitude of the motion and the phase constant under the assumption that at $t = 0$, the particle is moving in the region $x > 0$. Would your answers change if at $t = 0$, the particle is moving in the region $x < 0$?

3. A particle of mass m is dropped onto a horizontal surface which is elastic. The surface exerts a vertical force on the particle proportional to the distance which it is depressed in the vertical direction. Show that whilst it is in contact with the surface, the particle will perform simple harmonic motion. Find an expression for the constant of proportionality associated with the force exerted by the surface in terms of x_0, the distance which the surface is depressed when the particle is resting on it.

4. A particle, performing simple harmonic motion about a fixed origin O, passes points P_1 and P_2 at distances d_1 and d_2 from O moving towards O with speeds v_1 and v_2 respectively. Prove that the total energy per unit mass is given by

$$E' = \frac{d_1^2 v_2^2 - d_2^2 v_1^2}{2(d_1^2 - d_2^2)}.$$

5 A particle of mass m is suspended from a fixed point by a light spring of natural length l. The particle is in equilibrium when the length of the spring is l. The particle is pulled down a distance μl from its equilibrium position O and released from rest. Assuming Hooke's law find the frequency of the resulting oscillation and obtain an expression for the displacement $x(t)$ of the particle relative to O.

6 A light elastic string is stretched between two points, one lying vertically below the other. A particle is attached to the mid point of the string causing it to sink a distance l. Show that the period of small vertical oscillations of the particle about its equilibrium position is equal to $2\pi\sqrt{l/g}$.

3.3 Damped Oscillations

Most of the objects which surround us are at rest, usually in stable equilibrium. These objects are subject to many everyday disturbances and it seems surprising, therefore, that we do not observe a multitude of objects about us each oscillating about its position of equilibrium. When discussing simple harmonic oscillations in the last two sections no consideration was given to the effect that any drag forces might have on the oscillation. We shall see in this section that the oscillations are not permanent in the presence of such forces but die away with time so that objects, when disturbed in a resisting medium, always return to their positions of equilibrium.

Example 5

The particle in Example 1 oscillates whilst immersed in a beaker of oil, the resistance of the oil to the motion being modelled by a Stokes' drag. Write down the equation of motion of the particle.

SOLUTION

Stokes' drag D is given by

$$D = 3\pi\eta d|\dot{x}|,$$

$|\dot{x}|$ being the speed of the particle. The drag force acts in the opposite direction to the direction of motion and therefore has the opposite sign to the velocity $\dot{x}$. It follows that this drag force is $-3\pi\eta d\dot{x}$. Including this force on the right hand side of the equations of motion found in Example 1 leads to the equation

$$\ddot{x} + \frac{3\pi\eta d\dot{x}}{m} + \frac{\lambda}{ml_0}x = 0$$

$$\text{or} \quad \ddot{x} + \frac{3\pi\eta d\dot{x}}{m} + \frac{\lambda x}{ml_0} = g + \frac{\lambda}{m}$$

depending on whether x is taken relative to the position of equilibrium or to the point of suspension, respectively. The direction of the three forces acting on the particle are shown in Fig 3.6. ●

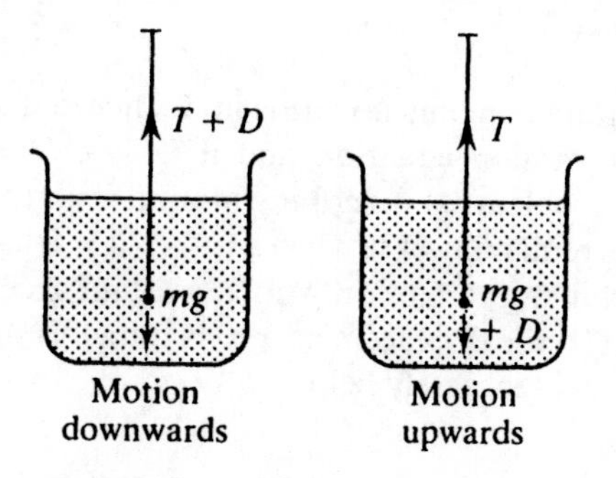

Fig 3.6 Oscillations in oil.

The above example shows that the presence of a drag force will modify the simple harmonic motion equation of an oscillating particle. Attention will be confined in this section to Stokes' drag so that the equation of motion is modified by the inclusion of a drag force proportional to the speed $|\dot{x}|$ of the particle. This force is always directed in the opposite sense to the velocity and it is convenient to write it as $-mk\dot{x}$ so that the mass m of the particle can be cancelled throughout the equation of motion. The constant k is positive. The simple harmonic motion equation is therefore replaced by

$$\ddot{x} + \omega^2 x = -k\dot{x}$$

$$\text{or} \quad \ddot{x} + k\dot{x} + \omega^2 x = 0.$$

This equation is called the **damped simple harmonic motion** equation and any mechanical system whose motion is determined by this equation is called a **damped simple harmonic oscillator**. The constant k is called the **damping constant** and $k\dot{x}$ is called the **damping term**. Two examples of a damped oscillator are the shock absorbers on cars, often called dampers, and the hydraulic door springs which are found on doors of public buildings.

The damped simple harmonic motion equation is another example of a second order linear homogeneous differential equation. If you know any standard techniques for solving such equations you should apply them to the equation. If not we shall have to resort to a manipulative trick, albeit a rather neat trick! Let us change the unknown from $x(t)$ to $X(t)$ defined by the equation

$$x = e^{-\frac{kt}{2}} X.$$

Differentiating gives

$$\dot{x} = e^{-\frac{kt}{2}}\dot{X} - \frac{k}{2}e^{-\frac{kt}{2}}X$$

$$\text{and} \quad \ddot{x} = e^{-\frac{kt}{2}}\ddot{X} - ke^{-\frac{kt}{2}}\dot{X} + \frac{k^2}{4}e^{-\frac{kt}{2}}X.$$

Substituting these into the damped simple harmonic motion equation and cancelling a common factor $e^{-\frac{kt}{2}}$ yields

$$\ddot{X} + (\omega^2 - \frac{k^2}{4})X = 0.$$

Notice that this equation contains no terms in $\dot{X}$. Indeed if $\omega^2 - k^2/4 > 0$ it is just the simple harmonic motion equation and if $\omega^2 - k^2/4 < 0$ it is the equation analyzed in Tutorial Problem 3.4 describing motion close to a position of unstable equilibrium. Even the remaining case when $\omega^2 - k^2/4 = 0$ has been met before, it is the equation corresponding to motion with a constant acceleration. Knowing the general solution for $X(t)$ the displacement $x(t)$ is found by multiplying by $e^{-\frac{kt}{2}}$. The three cases are discussed separately below.

Case 1 $\omega^2 - k^2/4 > 0$

Putting $\omega'^2 = \omega^2 - k^2/4$ the general solution of the damped simple harmonic equation can be written as

$$x = e^{-\frac{kt}{2}} A \sin(\omega' t + \delta).$$

The motion is oscillatory, the effect of the damping term being to reduce the frequency of oscillation from the undamped value $\omega/2\pi$ to $\omega'/2\pi$ and to introduce

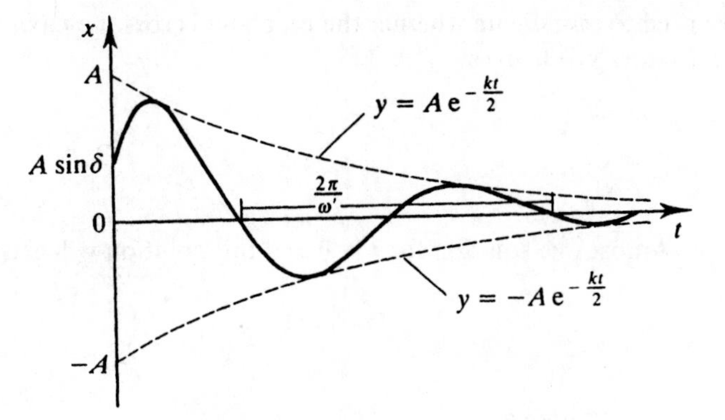

Fig 3.7 Weakly damped oscillator.

an exponentially decreasing amplitude $e^{-\frac{kt}{2}}A$. The graph of the displacement x against time t is given in Fig 3.7.

This case is called **weak damping** because the effect of the damping term is not sufficient to prevent oscillations.

Case 2 $\omega^2 - k^2/4 < 0$

Putting $\omega'^2 = -(\omega^2 - k^2/4)$ the general solution of the damped simple harmonic motion equation can be written as

$$x = e^{-\frac{kt}{2}}(c_1 e^{w't} + c_2 e^{-w't}).$$

The exponential function is non periodic and the motion is no longer oscillatory. The solution can also be written as

$$x = c_1 e^{(w' - \frac{k}{2})t} + c_2 e^{(-w' - \frac{k}{2})t}.$$

It follows from the definition of ω' that $\omega' < k/2$ and so both exponentials, and therefore the displacement x, will tend to zero as $t \to \infty$. This case is called **strong damping**. The graph of the displacement can be of two different forms and it is instructive to give the conditions which distinguish them in terms of some initial conditions rather than in terms of the arbitrary constants c_1 and c_2. Suppose that at time $t = 0$ the displacement of the oscillator is $x_0 > 0$ and the velocity v_0. Then

$$x_0 = c_1 + c_2$$

and $$v_0 = c_1(\omega' - \frac{k}{2}) + c_2(-\omega' - \frac{k}{2}).$$

Solving these equations gives

$$c_1 = \frac{1}{2\omega'}[v_0 + (\frac{k}{2} + \omega')x_0]$$

and $$c_2 = \frac{1}{2\omega'}[-v_0 + (-\frac{k}{2} + \omega')x_0].$$

Subsequently the oscillator – this is now something of a misnomer! – will tend to its equilibrium position $x = 0$ as $t \to \infty$. In order to sketch the graph of the displacement we need to investigate whether the graph will cross the t axis for any finite time $t > 0$. Putting $x = 0$ gives

$$c_1 e^{(\omega' - \frac{k}{2})t} + c_2 e^{(-\omega' - \frac{k}{2})t} = 0$$

$$\text{so that} \quad e^{2\omega' t} = -\frac{c_2}{c_1}.$$

This equation has at most one solution for $t > 0$ and this solution will exist if and only if

$$-\frac{c_2}{c_1} > 1.$$

This inequality can be rewritten as

$$-\frac{c_2}{c_1} > \frac{c_1}{c_1} \quad \text{or} \quad \frac{c_1 + c_2}{c_1} < 0.$$

Now $c_1 + c_2 = x_0$ which is being taken to be positive. It follows that the graph of the displacement x against time will cut the positive t axis if and only if

$$c_1 < 0$$

$$\text{i.e.} \quad v_0 < -(\frac{k}{2} + \omega')x_0.$$

The graph will take one of the two forms illustrated in Fig 3.8.

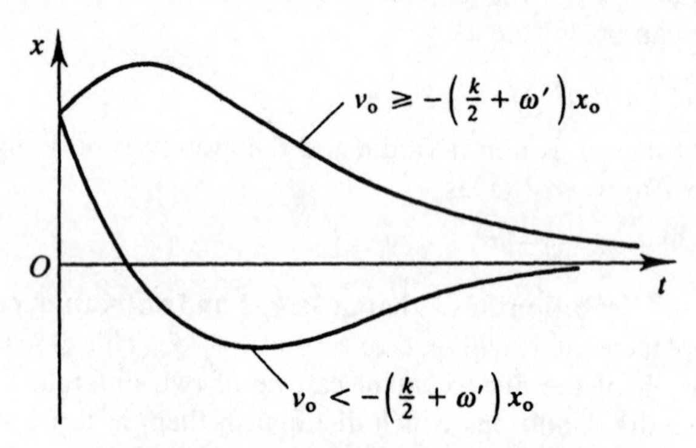

Fig 3.8 Strongly damped oscillator.

Case 3 $\omega^2 - k^2/4 = 0$

The general solution of the damped simple harmonic motion equation can be written as

$$x = e^{-\frac{kt}{2}}(c_1 t + c_2).$$

The motion is again non oscillatory and $x \to 0$ as $t \to \infty$. As in the last case, suppose that at time $t = 0$ the displacement of the oscillator is $x_0 > 0$ and the velocity v_0. Then

$$x_0 = c_2 \quad \text{and} \quad v_0 = -\frac{k}{2}c_2 + c_1.$$

Eliminating c_2 between these equations gives

$$c_1 = v_0 + \frac{k}{2}x_0.$$

When $x = 0$ the time t is given by

$$t = -\frac{c_2}{c_1}$$

and so the graph of the displacement will cut the positive t axis if and only if

$$-\frac{c_2}{c_1} > 0.$$

Now $c_2 = x_0 > 0$ and so this inequality becomes

$$c_1 < 0$$

$$\text{or} \quad v_0 < -\frac{k}{2}x_0.$$

It follows that the graph of the displacement against time is as in Fig 3.8, the conditions for the different forms being obtained by putting $\omega' = 0$. This case is called **critical damping.**

The frequency $\omega/2\pi$ of the undamped oscillator is called the **natural frequency**, to distinguish it from the frequency of the damped oscillations which occur in the case of weak damping. In the case of strong or critical damping the oscillator is said to **overshoot** if it passes its equilibrium position before tending to it as $t \to \infty$. In practical applications of the theory, for example in the design of the shock absorbers used to dampen the springs of a car, it is usual to require the oscillator to return towards its equilibrium position without overshooting and as quickly as possible. Of course the oscillator never quite reaches its equilibrium position. Denote the displacement in the cases of critical and strong damping by x_c and x_s, respectively. Then, assuming that in both cases $x = x_0 > 0$ and $\dot{x} = v_0$ when $t = 0$, the ratio

$$\frac{x_c}{x_s} = \frac{(v_0 + \frac{k}{2}x_0)t + x_0}{\frac{1}{2\omega^2}[v_0 + (\frac{k}{2} + \omega')x_0]e^{\omega' t} + \frac{1}{2\omega'}[-v_0 + (\frac{k}{2} + \omega')x_0]e^{-\omega' t}}.$$

As $t \to \infty$ the limit of this ratio can be found using L'Hopital's rule or by observing that the exponential $e^{\omega' t}$ in the denominator tends to infinity quicker than the numerator. Hence

$$\frac{x_c}{x_s} \to 0$$

and therefore the critically damped oscillator returns towards its equilibrium quicker than the strongly damped oscillator. For this reason the engineer usually designs damping mechanisms to give critical damping.

Example 6

An experiment is set up to determine the damping constant k of a weakly damped oscillator by observing the frequency of the oscillation ω' and two consecutive positions of maximum displacement, x_1 and x_2. Prove that

$$k = \frac{\omega'}{\pi}\log(\frac{x_1}{x_2}).$$

SOLUTION

The displacement x of the weakly damped oscillator at time t is given by

$$x = e^{-\frac{kt}{2}}A\sin(\omega' t + \delta).$$

At maximum or minimum displacement the velocity $\dot{x}$ must be zero. Hence the corresponding time satisfies

$$e^{-\frac{kt}{2}}A\omega'\cos(\omega' t + \delta) - \frac{k}{2}e^{-\frac{kt}{2}}A\sin(\omega' t + \delta) = 0$$

$$\text{or} \quad \tan(\omega' t + \delta) = \frac{2\omega'}{k}.$$

The times t_2 and t_1 of two consecutive positions of maximum displacement will therefore be related by $t_2 - t_1 = 2\pi/\omega'$; note that $t_2 - t_1 = \pi/\omega'$ would correspond to consecutive positions of maximum/minimum displacement. Now

$$x_1 = e^{-\frac{kt_1}{2}}A\sin(\omega' t_1 + \delta)$$

$$\text{and} \quad x_2 = e^{-\frac{kt_2}{2}}A\sin(\omega' t_2 + \delta) = e^{-\frac{kt_2}{2}}A\sin(\omega' t_1 + \delta).$$

Hence

$$\frac{x_1}{x_2} = \frac{e^{-\frac{kt_1}{2}}}{e^{-\frac{kt_2}{2}}} = e^{\frac{k}{2}(t_2 - t_1)} = e^{\frac{k\pi}{\omega'}}.$$

From this

$$k = \frac{\omega'}{\pi}\log(\frac{x_1}{x_2}).$$

EXERCISES ON 3.3

1. The energy E' of a simple harmonic oscillator is defined by

$$E' = \tfrac{1}{2}\dot{x}^2 + \tfrac{1}{2}\omega^2 x^2.$$

Show that if the oscillator is damped then E' decreases at a rate proportional to the square of the speed of the oscillator.

2. A critically damped simple harmonic oscillator ($k = 2\omega$) starts from rest at displacement $x_0 > 0$. Show that x is always decreasing with time. Prove that the greatest speed occurs when $x = 2x_0/e$.

3. Prove that a critically damped simple harmonic oscillator cannot overshoot if its initial acceleration is negative, it being assumed that its initial displacement is positive. Investigate whether the same is true of a strongly damped oscillator.

4. The critically damped oscillator

$$\ddot{x} + k\dot{x} + \omega^2 x = 0,$$

with $k = 2\omega$, has displacement $x = x_0$ and velocity $\dot{x} = v_0$ at time $t = 0$. At time $t = T$ the oscillator passes through its position of equilibrium $x = 0$. Find an expression for the damping constant k in terms of x_0, v_0 and T. Show also that the oscillator never subsequently passes through $x = 0$.

5. A strongly damped harmonic oscillator passes its equilibrium position O and at time T later it reaches its position of maximum distance from O. Show that T is given by

$$2\omega' T = \log_e \lambda\left(\frac{k/2+\omega'}{k/2-\omega'}\right).$$

6. A strongly damped oscillator is described by

$$\theta = e^{-\frac{kt}{2}}[ae^{-\frac{k't}{2}} + be^{\frac{k't}{2}}],$$

where $k'^2 = k^2 - 4\omega^2$. The oscillator starts from rest at $\theta = \theta_0$. Show that $a = \frac{k'-k}{2k'}\theta_0$ and $b = \frac{k'+k}{2k'}\theta_0$. Deduce that the oscillator passes the equilibrium position $\theta = 0$ once before finally tending to it.

3.4 Forced Damped Oscillations

A discussion of simple harmonic motion cannot be complete without introducing a sinusoidal **driving force** term onto the right hand side of the damped simple harmonic motion equation to give

$$\ddot{x} + k\dot{x} + \omega^2 x = F \sin pt.$$

F and p are the amplitude and angular frequency of the driving force, respectively. This equation is called the **forced, damped simple harmonic motion equation** and any mechanical system whose motion is determined by this equation is called a **forced damped simple harmonic oscillator**.

Example

The point of suspension in Example 5 is made to oscillate vertically about a fixed point O with amplitude a and frequency $p/2\pi$ so that its displacement relative to O is $a \sin pt$. Obtain the equation of motion in terms of the displacement x of the particle relative to O.

SOLUTION

In the absence of the Stokes' drag the equation of motion of the particle would be almost as in part (ii) of the solution to Example 1, the only difference being that the extension of the spring is now given by $x - a \sin pt - l_0$ so that

$$m\ddot{x} = mg - \frac{\lambda}{l_0}(x - a \sin pt - l_0).$$

Adding in the Stokes' drag gives

$$m\ddot{x} = mg - \frac{\lambda}{l_0}(x - a \sin pt - l_0) - 3\pi\eta d\dot{x}$$

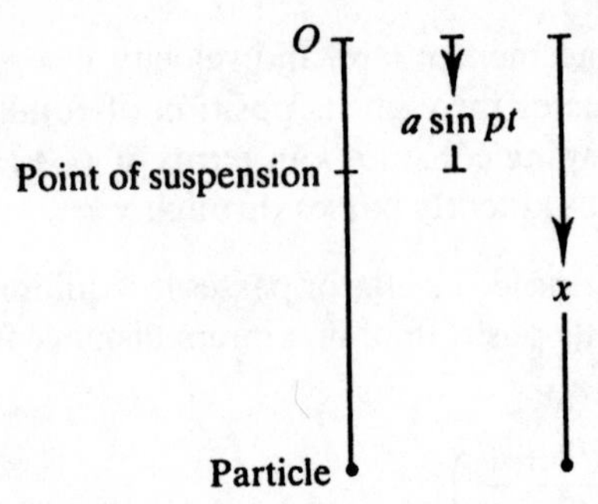

Fig 3.9 Oscillating point of suspension.

which can be written as

$$\ddot{x} + \frac{3\pi\eta d}{m}\dot{x} + \frac{\lambda x}{ml_0} = \frac{\lambda a}{ml_0}\sin pt + g + \frac{\lambda}{m}.$$

Notice that the position of equilibrium is no longer fixed and therefore cannot be chosen as the origin as was done in part (i) of the solution to Example 1. Nevertheless the constant terms on the right hand side of the equation of motion can be eliminated by choosing a new origin O' for which

$$x = x' + \frac{ml_0 g}{\lambda} + l_0.$$

This origin coincides with the position of equilibrium which would arise if the point of suspension were fixed at O. Dropping the prime on x' the equation of motion then takes the form

$$\ddot{x} + \frac{3\pi\eta d}{m}\dot{x} + \frac{\lambda}{ml_0}x = \frac{\lambda a}{ml_0}\sin pt.$$

TUTORIAL PROBLEM 3.5

Load a spring with masses so that, whilst it is not at the point of being deformed by the loading, it is nevertheless considerably extended when held vertically. For each of the experiments below, you should try to predict what will happen before you carry it out and then, after noting what actually happens, try to explain the phenomenon observed.

Move the hand holding the spring vertically up and down, gradually increasing the frequency of the oscillation of your hand and note what happens.

Now take the loaded spring, hold it vertically again and this time move the hand holding the spring up and down very quickly indeed. Again note what happens. Repeat this experiment, but this time gradually decrease the frequency of oscillation of the hand. Note what happens.

What happens in the above experiments if you vary the mass or use a different spring? Later in this section, you should be able to justify your answers in terms of the mathematics of the situations.

The forced damped simple harmonic motion equation is a second order linear inhomogeneous differential equation and again you may know standard methods for solving such equations. The presence of the term $F \sin pt$ on the right hand side suggests that a particular solution $x_0(t)$ of the equation might exist of the form

$$x_0 = \alpha \sin pt + \beta \cos pt.$$

Differentiating gives

$$\dot{x}_0 = \alpha p \cos pt - \beta p \sin pt$$
$$\text{and} \quad \ddot{x}_0 = -\alpha p^2 \sin pt - \beta p^2 \cos pt.$$

Hence

$$\ddot{x}_0 + k\dot{x}_0 + \omega^2 x_0 = [\alpha(\omega^2 - p^2) - \beta kp] \sin pt + [\alpha kp + \beta(\omega^2 - p^2)] \cos pt.$$

It follows that x_0 satisfies the equation if and only if

$$\alpha(\omega^2 - p^2) - \beta kp = F$$
$$\text{and} \quad \alpha kp + \beta(\omega^2 - p^2) = 0.$$

From these α and β can be determined to yield the solution

$$\begin{aligned} x_0 &= \frac{F}{(\omega^2 - p^2)^2 + p^2k^2}[(\omega^2 - p^2) \sin pt - pk \cos pt] \\ &= \frac{F}{(\omega^2 - p^2) + p^2k^2}[(\omega^2 - p^2)^2 + p^2k^2]^{\frac{1}{2}} \sin(pt + \delta) \\ &= \frac{F}{[(\omega^2 - p^2)^2 + p^2k^2]^{\frac{1}{2}}} \end{aligned}$$

This particular solution of the equation represents an oscillation with frequency equal to the frequency of the forcing term and with amplitude $F/[(\omega^2 - p^2)^2 + p^2k^2]^{\frac{1}{2}}$. If $x(t)$ is any other solution of the equation then

$$\ddot{x} + k\dot{x} + \omega^2 x = F \sin pt$$
$$\text{and} \quad \ddot{x}_0 + k\dot{x}_0 + \omega^2 x_0 = F \sin pt$$

so that the difference $X(t) = x(t) - x_0(t)$ satisfies

$$\ddot{X} + k\dot{X} + \omega^2 X = 0.$$

It follows that the general solution of the forced damped simple harmonic motion equation is obtained by adding $x_0(t)$ onto the general solution of the unforced damped simple harmonic motion equation. We saw in the last section that in all cases the general solution to the unforced equation tends to zero as t tends to infinity; the corresponding component of the motion of the forced oscillator is called the **transient** motion. As this transient motion becomes negligible the motion of the oscillator is determined by the particular solution $x_0(t)$ of the forced equation, this is known as the **steady state motion**. Hence for large times t the motion becomes oscillatory with frequency equal to the frequency of the forcing term.

For an undamped forced oscillator, $k = 0$, and the amplitude of the steady state oscillations becomes

$$\frac{F}{|\omega^2 - p^2|}.$$

If the frequency $p/2\pi$ of the driving force tends to the natural frequency $\omega/2\pi$ of the oscillator this amplitude will tend to infinity. This phenomenon is known as **resonance**. In the design of any structure it is essential to ensure that the natural frequency of the structure is not equal to the frequency of any disturbance which might affect the structure. If this is not done, very serious consequences will follow. You may have seen films of the Tacoma Bridge disaster or have experienced resonant oscillations of a car as it travels over regularly spaced bumps in the road. You should have observed resonant behaviour when carrying out the experiments in Tutorial Problem 3.5.

EXERCISES ON 3.4

1. A simple harmonic oscillator whose period in vacuum is 4s is placed in a medium which resists the motion with a force proportional to the velocity. The amplitude of the oscillator decreases by a factor 2 on each complete swing. Find the period of oscillation. The oscillator is acted upon in turn by sinusoidal forces with periods of 3s and 5s and with equal amplitudes. Find the ratios of the amplitudes of the steady state oscillations.

2. Discuss the effect of immersing a particle, performing simple harmonic motion on the end of a vertical spring, in a beaker of fluid which is also performing simple harmonic motion in a vertical direction. You may assume that the drag force of the fluid on the particle is proportional to the velocity of the particle relative to the fluid.

3. Resonance occurs when the frequency of the driving force acting on an undamped oscillator tends to the natural frequency of the oscillator. If these frequencies are equal the equation of motion becomes

$$\ddot{x} + \omega^2 x = F \sin \omega t.$$

   Solve this equation and show that the amplitude of the unbounded motion increases linearly with time.

Summary

- the **simple harmonic motion** equation is

$$\ddot{x} + \omega^2 x = 0$$

- the general solution of this equation can be written as

$$x = c_1 \sin \omega t + c_2 \cos \omega t \quad \text{or} \quad x = A \sin(\omega t + \delta)$$

- A is the **amplitude**, $\omega/2\pi$ the **frequency**, $2\pi/\omega$ the **period** or **periodic time** and δ the **phase constant** of the oscillation

- the energy equation of the oscillator is $\frac{1}{2}\dot{x}^2 + \frac{1}{2}\omega^2 x^2 = E'$ and $E' = \frac{1}{2}\omega^2 A^2$
- the motion of a particle moving close to a position of stable equilibrium is simple harmonic with $\omega^2 = \frac{1}{m}\frac{d^2V}{dx^2}\Big|_0$, where $V(x)$ is the potential energy of the particle and $\Big|_0$ denotes evaluation at the position of equilibrium
- the displacement of a particle moving close to a position of unstable equilibrium satisfies the equation

$$\ddot{x} - \omega^2 x = 0,$$

- where $\omega^2 = -\frac{1}{m}\frac{d^2V}{dx^2}\Big|_0$; the general solution of this equation is

$$x = c_1 e^{\omega t} + c_2 e^{-\omega t}$$

- the **damped simple harmonic motion equation** is

$$\ddot{x} + k\dot{x} + \omega^2 x = 0$$

- the general solution of this equation takes one of the following forms:

  weak damping (with $\omega^2 - k^2/4 = \omega'^2 > 0$) :

$$x = e^{-\frac{kt}{2}} A \sin(\omega' t + \delta)$$

  strong damping (with $\omega^2 - k^2/4 = -\omega'^2 < 0$):

$$x = e^{-\frac{kt}{2}}(c_1 e^{\omega' t} + c_2 e^{-\omega' t})$$

  critical damping (with $\omega^2 - k^2/4 = 0$):

$$x = e^{-\frac{kt}{2}}(c_1 t + c_2)$$

- the motion of a damped oscillator is oscillatory only in the case of weak damping and then the **frequency of oscillation** is $\omega'/2\pi$ and the **natural frequency** is $\omega/2\pi$
- in all cases the damped oscillator tends towards its equilibrium position as $t \to \infty$
- **overshoot** occurs in the case of strong and critical damping when the particle passes its equilibrium position before tending to it as $t \to \infty$
- the **forced damped simple harmonic motion equation** is

$$\ddot{x} + k\dot{x} + \omega^2 x = F \sin pt$$

- the general solution of this equation is obtained by adding the particular solution

$$\frac{F}{(\omega^2 - p^2)^2 + p^2k^2}[(\omega^2 - p^2)\sin pt - pk \cos pt],$$

  representing an oscillation with frequency $p/2\pi$, to the appropriate form of the general solution of the corresponding unforced equation
- the **transient** motion of a forced oscillator corresponds to the general solution of the unforced equation and the **steady state** motion corresponds to the above particular solution of the forced equation

- **resonance** of an undamped oscillator occurs when the amplitude of the steady state motion tends to infinity as the frequency of the driving force tends to the natural frequency of the oscillator.

FURTHER EXERCISES

1. A heavy particle is attached to the midpoint of a light elastic string which is then stretched between two fixed points lying in a vertical line. Assuming that each portion of the string remains stretched and that the tensions in the string obey Hooke's law, prove that the frequency of vertical oscillations of the particle about its position of equilibrium is the same as the frequency of oscillations the particle would perform if oscillating vertically at one end of a light elastic string having the same modulus of elasticity as the first string but having one quarter of its natural length. Here the other end of the second elastic string is assumed to be attached to a fixed point.

2. A horizontal plate is moving vertically such that its displacement from a fixed origin at time t is given by $x = a \sin \omega t$, where x is measured positively upwards. A particle of mass m is resting on the plate. What is the maximum value of the amplitude a, if the particle is to remain in contact with the plate.

3. Show that the maximal displacements of a weakly damped simple harmonic oscillator are in geometric progression.

4. Show that a critically damped oscillator can overshoot its equilibrium position $x = 0$ no more than once and that if it does so then the inverse of the damping constant k is equal to half the time taken for the oscillator to attain its maximum displacement after passing its equilibrium position.

5. A particle of mass m is attached to a horizontal spring, as shown in the diagram, which is initially unstretched and of natural length l_0. The table on which the

Fig 3.10 Particle on horizontal spring.

particle rests is rough and exerts a constant horizontal frictional force of mg on the particle in such a direction as to oppose the motion of the particle. The particle is given an initial speed so as to extend the spring. Write down an equation of motion which applies from the start of the motion until the speed is first zero. Solve this equation. The speed of the particle is first zero when $\omega t = \pi/4$, where $\omega^2 = \lambda/ml_0$. Show that the tension is then less than mg so that the particle remains at rest.

6. Write down expressions for the displacements of the transient and steady state motions of the forced damped oscillator

$$\ddot{x} + k\dot{x} + \omega^2 x = F \sin pt$$

in the case of weak damping ($k^2 < 4\omega^2$). In the absence of the forcing term the oscillator is observed to oscillate with a period of 6s, the amplitude of the oscillation decreasing by a factor 10 on each complete oscillation. Prove that the natural period of the oscillator is approximately 5.7s. The oscillator is acted on by a sinusoidal forcing term with period 8s and amplitude 0.5ms^{-1}. Show that the amplitude of the steady state oscillations is 0.7m.

7. The forced undamped simple harmonic equation is written in the form

$$\ddot{x} + \omega^2 x = F \sin(\omega + \epsilon)t.$$

Prove that if $x = \dot{x} = 0$ when $t = 0$, the solution of this equation can be written as

$$x = \frac{F}{\omega^2 - (\omega + \epsilon)^2}\left[-\frac{(\omega + \epsilon)}{\omega}\sin \omega t + \sin(\omega + \epsilon)t\right].$$

Show that this solution approximates to the form

$$x = \frac{2F}{\omega^2 - (\omega + \epsilon)^2}\sin \frac{\epsilon}{2}t \cos(\omega + \frac{\epsilon}{2})t$$

when the frequency of the forcing term is very close to the natural frequency of the oscillator, ie when $\epsilon \ll \omega$. Noting that the subsequent motion can be considered as an oscillation with a slowly varying amplitude of $2F|\sin \frac{\epsilon}{2}t|/|\omega^2 - (\omega + \epsilon)^2|$, or otherwise, sketch the graph of x against t.

4 • Motion in Space

Many of the ideas discussed in Chapter 2 are generalized to the motion of a particle moving in space. A vectorial approach is used throughout, coordinates, when used, being adapted to the problem being discussed. The importance of the choice of frame of reference used to describe the motion of a given particle is stressed and the role played by Newton's first law of motion in the choice of the frame of reference is discussed. Three different types of conservation laws are derived. These are important in that they provide first integrals of the equation of motion which can be written down without the need for carrying out actual integrations. Newton's third law of motion is stated in both a weak and strong form. Much of this Chapter is concerned with development of the theory and the opportunity for tackling exercises is therefore somewhat restricted.

4.1 Kinematics of a Particle Moving in Space

Consider a particle moving in space. The definitions and results which we are about to discuss are generalizations of those discussed in Section 2.1 for a particle P moving on a straight line. In Section 2.1 the position of P was specified, relative to a given origin O lying on the line of motion, by a directed number x, the displacement of P relative to O. You will have noticed the fact that if the line of motion is chosen as the x-axis of a cartesian coordinate system with origin O, then the displacement of P relative to O is just the x coordinate of P. An obvious generalization for a particle P moving in space would be to specify the position of P relative to a given origin by the cartesian coordinates (x,y,z) of P relative to a given set of axes passing through O. In order to simplify later calculations we confine attention here to right handed rectangular cartesian coordinates for which the axes are mutually perpendicular and orientated with Oz in the direction of the thumb of the right hand when the curled fingers of that hand are scooped from Ox to Oy, as in Fig 4.1. The axes are said to form a **frame of reference** or **frame** against which the motion of the particle is viewed.

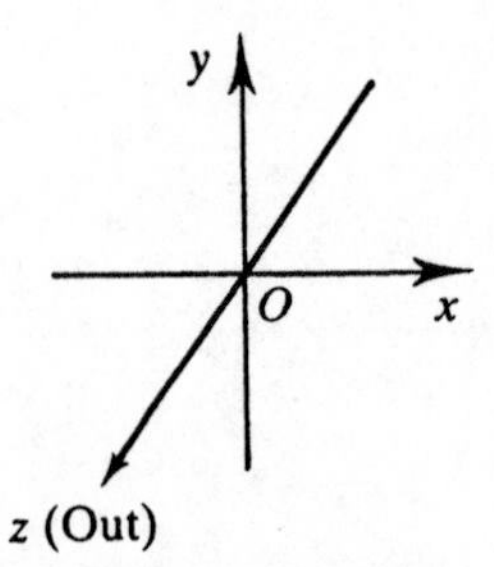

Fig 4.1 Right handed rectangular cartesian axes.

Cartesian coordinates are not always the most convenient coordinates to use when discussing particular motions. For example, it is far easier to use plane polar coordinates and the equation $r = a$ to specify motion on a circle of radius a centred at O than to use cartesian coordinates and the equation $x^2 + y^2 = a^2$. For this reason we prefer to adopt a vectorial approach when discussing the general motion of a particle P in space. In this approach the position of P is specified by the displacement vector $\overrightarrow{OP}$ relative to a given origin O. This is an alternative generalization of the displacement x; for straight line motion the displacement is indeed a vector, its sign indicating the direction of the vector. The displacement vector $\overrightarrow{OP}$ is called the **position vector of P relative to O** and is denoted by $\mathbf{r}$. This vectorial approach is closely related to the coordinate approach. Taking a right handed set of orthonormal basis vectors $\hat{\mathbf{i}}, \hat{\mathbf{j}}$ and $\hat{\mathbf{k}}$, the components of the position vector $\mathbf{r}$ of P relative to O are just the cartesian coordinates (x, y, z) of P relative to axes drawn parallel to the basis vectors and passing through O, so that

$$\mathbf{r} = x\hat{\mathbf{i}} + y\hat{\mathbf{j}} + z\hat{\mathbf{k}}.$$

In this context we refer to $\hat{\mathbf{i}}, \hat{\mathbf{j}}, \hat{\mathbf{k}}$ as a **(rectangular) cartesian basis** and the components (v_x, v_y, v_z) of a vector $\mathbf{v}$ relative to such a basis as the **(rectangular) cartesian components of the vector**. The word rectangular will normally be omitted because, as has already been stated, only such systems are used here.

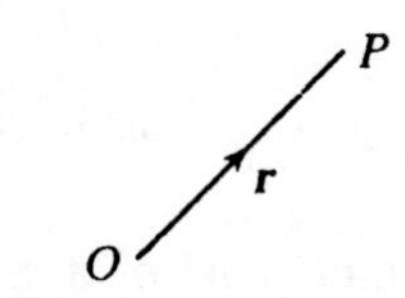

Fig 4.2 The position vector.

Two simple results follow directly from Fig 4.2, namely

- the position vector of a point relative to itself is the zero vector
- the position vector of O relative to P is minus the position vector of P relative to O.

The position vector of a given point depends on the choice of origin. Let $\mathbf{r}$ and $\mathbf{r}'$ be the position vectors of P relative to two different origins O and O', respectively.

From the vector triangle in Fig 4.3 it follows that

$$\overrightarrow{O'P} = \overrightarrow{O'O} + \overrightarrow{OP}$$

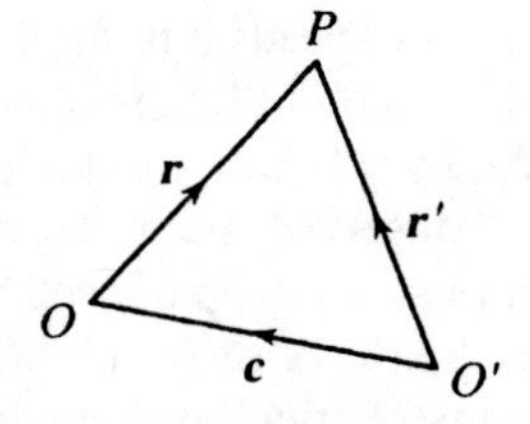

Fig 4.3 Change of origin.

so that, interchanging the vectors on the right hand side,

$$\mathbf{r}' = \mathbf{r} + \mathbf{c},$$

where $\mathbf{c}$ is the position vector of O relative to O'. You may like to think of $\mathbf{c}$ as a **connecting vector**, connecting the second origin O' to the first origin O. Notice the juxtaposition of the two O's on the right hand side of the first equation. The second equation generalizes the result

$$\text{disp of } P \text{ rel to } O' = \text{disp of } P \text{ rel to } O + \text{disp of } O \text{ rel to } O',$$

obtained in Section 2.1 for motion on a straight line, to motion in space, provided that disp is now taken to be an abreviation for the displacement vector or position vector. Notice that the proof of the result given here is far easier than the proof in Tutorial Problem 2.1 for the case of motion on a straight line; this illustrates the power of the vectorial approach.

TUTORIAL PROBLEM 4.1

Review together your knowledge of vector algebra, in particular:

(i)the products of two or more vectors with their geometrical interpretations and consequences

(ii)the identity for $\mathbf{u} \times (\mathbf{v} \times \mathbf{w})$

(iii)the calculation of sums and products using components relative to a right handed orthonormal basis

(iv)the determination of the components of a given vector relative to a right handed orthonormal basis

(v) the vector equations of a straight line and of a plane.

What do you know about the calculus of vector valued functions?

The observed path of a moving particle depends both on the observer and on the frame of reference used by the observer. To illustrate this, consider first the path of a raindrop as seen by an observer O at the side of a railway track and an observer O' sitting in a train. If there is no wind the observer O will see the path of the raindrop to be a vertical straight line. This is of course also the path seen by the observer O' and traced out by a raindrop running down the window of the train, provided that the train is stationary. You will have noticed that when the train picks up speed the straight line path traced out by the raindrop is no longer vertical. In fact you might also have noticed that the angle between this straight line and the vertical is small for low speeds but large for high speeds. As the train accelerates the slope of the path traced out by the raindrop must change so that the path of the raindrop as seen by O' can no longer be a straight line. This demonstrates clearly that the observed path depends on the motion of the observer, the straight line path of a raindrop as seen by the observer O at the side of the track is curved when seen by the observer O' sitting in the accelerating train. Now consider the motion of an insect crawling from the centre O of a record, along a radial straight line. If an observer at O uses a frame of reference fixed to the turntable then even when the record is rotating the path of the insect will be seen to be a radial straight line. However if the same observer uses a frame of reference fixed to the cabinet of the turntable then the path of the insect will be seen to be a spiral. This demonstrates that the observed path depends on the motion of the frame of reference, the straight line path viewed against one frame of reference becomes a spiral when viewed against a frame of reference rotating relative to the first.

The above discussion indicates that it is essential to specify both the observer and the frame of reference used by the observer when describing the motion of a particle. The frame of reference will be denoted by S and in what follows the words "relative to the observer O" will be taken to include not only the position O of the observer but also the frame of reference S used by the observer.

TUTORIAL PROBLEM 4.2

Why was the concept of the frame of reference not introduced when discussing motion on a straight line?

The position vector $\mathbf{r}$ of a moving particle P relative to a given observer O will be a function of time,

$$\mathbf{r} = \mathbf{r}(t).$$

Knowing this function, the cartesian coordinates of P relative to the axes which define the frame of reference S used by the observer can be calculated for different values of the time t and the corresponding positions of P plotted. Joining the resulting points by a smooth curve will give the path of the particle relative to the observer O. The function $\mathbf{r}(t)$ is a parametric representation of this path and when this function is given, the equation $\mathbf{r} = \mathbf{r}(t)$ is referred to as the equation of the path. The derivative of $\mathbf{r}(t)$ is defined to be the **velocity of *P* relative to *O*** and is written as either $\dot{\mathbf{r}}$ or $\frac{d\mathbf{r}}{dt}$. You will remember that the derivative of a vector valued function is the vector whose cartesian components are the derivatives of the cartesian components of the vector being differentiated. Again the frame of reference S is important here because it specifies the cartesian axes and cartesian basis to be used by the observer in calculating the derivative. Since

$$\mathbf{r} = x\hat{\mathbf{i}} + y\hat{\mathbf{j}} + z\hat{\mathbf{k}}$$

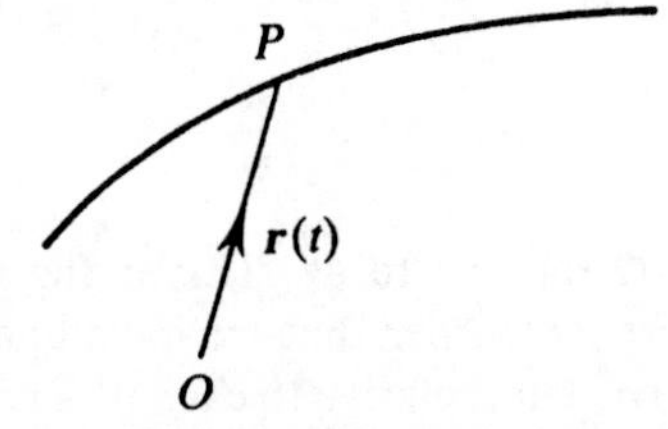

Fig 4.4 Path of the particle relative to O.

it follows that the velocity $\mathbf{v}$ is given by

$$\mathbf{v} = \frac{dx}{dt}\hat{\mathbf{i}} + \frac{dy}{dt}\hat{\mathbf{j}} + \frac{dz}{dt}\hat{\mathbf{k}} = \dot{x}\hat{\mathbf{i}} + \dot{y}\hat{\mathbf{j}} + \dot{z}\hat{\mathbf{k}}.$$

The addition law of relative velocities,

$$\text{vel of } P \text{ rel to } O' = \text{vel of } P \text{ rel to } O + \text{vel of } O \text{ rel to } O',$$

obtained in Section 2.1 for motion on a straight line can be generalized to motion in space by differentiating $\mathbf{r}' = \mathbf{r} + \mathbf{c}$. Some care has to be taken because, as can be seen from Fig 4.3, $\mathbf{r}'$ and $\mathbf{c}$ are position vectors relative to the observer O' whereas $\mathbf{r}$ is a position vector relative to the observer O. These observers may use different frames of reference S and S' with different cartesian bases $\hat{\mathbf{i}}, \hat{\mathbf{j}}, \hat{\mathbf{k}}$ and $\hat{\mathbf{i}}', \hat{\mathbf{j}}', \hat{\mathbf{k}}'$. Writing

$$\mathbf{r} = x\hat{\mathbf{i}} + y\hat{\mathbf{j}} + z\hat{\mathbf{k}} \quad \text{and} \quad \mathbf{r}' = x'\hat{\mathbf{i}}' + y'\hat{\mathbf{j}}' + z\hat{\mathbf{k}}'$$

the velocity of P relative to O is given by

$$\mathbf{v} = \frac{dx}{dt}\hat{\mathbf{i}} + \frac{dy}{dt}\hat{\mathbf{j}} + \frac{dz}{dt}\hat{\mathbf{k}}$$

whereas the velocity of P relative to O' is given by

$$\mathbf{v}' = \frac{dx'}{dt'}\hat{\mathbf{i}}' + \frac{dy'}{dt'}\hat{\mathbf{j}}' + \frac{dz'}{dt'}\hat{\mathbf{k}}'.$$

As in Section 2.1 we will assume that the two observers use identical synchronized clocks so that $t' = t$ and

$$\mathbf{v}' = \frac{dx'}{dt}\hat{\mathbf{i}} + \frac{dy'}{dt}\hat{\mathbf{j}}' + \frac{dz'}{dt}\hat{\mathbf{k}}'.$$

The equation $\mathbf{r}' = \mathbf{r} + \mathbf{c}$ can be written as

$$x'\hat{\mathbf{i}}' + y'b\hat{\mathbf{j}}' + z'\hat{\mathbf{k}}' = x\hat{\mathbf{i}} + y\hat{\mathbf{j}} + z\hat{\mathbf{k}} + c_{x'}\hat{\mathbf{i}}' + c_{y'}\hat{\mathbf{j}}' + c_{z'}\hat{\mathbf{k}}'$$

and before differentiating this equation with respect to time t we must decide which frame of reference to use in order to calculate this derivative. Let us use the frame S', in which case it is essential to realize that it will be necessary to differentiate $\hat{\mathbf{i}}, \hat{\mathbf{j}}$ and $\hat{\mathbf{k}}$ with respect to time. This is done by differentiating the components of these vectors relative to the cartesian basis in the frame S'. If the frame S' is not rotating with respect to S (this does not preclude S' from being related to S by a fixed rotation) these components will all be constant and then the derivatives of $\hat{\mathbf{i}}, \hat{\mathbf{j}}$ and $\hat{\mathbf{k}}$ will all be zero. It then follows that

$$\frac{dx'}{dt}\hat{\mathbf{i}}' + \frac{dy'}{dt}\hat{\mathbf{j}}' + \frac{dz'}{dt}\hat{\mathbf{k}}' = \frac{dx}{dt}\hat{\mathbf{i}} + \frac{dy}{dt}\hat{\mathbf{j}} + \frac{dz}{dt}\hat{\mathbf{k}} + \frac{dc_{x'}}{dt}\hat{\mathbf{i}}' + \frac{dc_{y'}}{dt}\hat{\mathbf{j}}' + \frac{dc_{z'}}{dt}\hat{\mathbf{k}}'$$

so that

$$\mathbf{v}' = \mathbf{v} + \mathbf{V},$$

where $\mathbf{V}$ is the velocity of O relative to O'. This is the required addition law of relative velocities. You must remember that we have had to assume that the two frames of reference are not rotating relative to each other. With this assumption we now know from the above that we could have differentiated the equation

$$\mathbf{r}' = \mathbf{r} + \mathbf{c}$$

directly to give

$$\dot{\mathbf{r}}' = \dot{\mathbf{r}} + \dot{\mathbf{c}}$$

which can be rewritten in the form $\mathbf{v}' = \mathbf{v} + \mathbf{V}$ obtained above. There are many applications of the addition law of relative velocities, for example in aviation it is used to evaluate the ground velocity of an aeroplane, knowing the wind velocity and the air velocity of the aeroplane.

In the above derivation and in Section 2.1 we have assumed that the observers O and O' use identical clocks which have been synchronized and which therefore remain synchronized because of the Newtonian assumption of the universality of time. Then $t' = t$ and derivatives with respect to t' are identical to derivatives with respect to t. This enabled us to differentiate each term of the equation $\mathbf{r}' = \mathbf{r} + \mathbf{c}$ with respect to t to obtain the addition law of relative velocities. Suppose that the identical clocks are not synchronized so that $t' = t +$ constant, the value of the constant indicating how "fast" or "slow" the reading of one clock is relative to the other. Using the chain rule

$$\frac{d}{dt} = \frac{dt'}{dt}\frac{d}{dt'} = \frac{d}{dt'}$$

and therefore derivatives with respect to t' are again identical to derivatives with respect to t. It follows that we do not need to require the observers to synchronize their identical clocks.

In order to obtain an interpretation of the velocity $\mathbf{v}$, consider a fixed point P_0 lying on the path of the particle P and suppose that the observer's clock is set to zero as the particle passes this point. From Fig 4.5 it follows that

$$\mathbf{r}'(t) = \mathbf{r}(t) - \mathbf{r}(0).$$

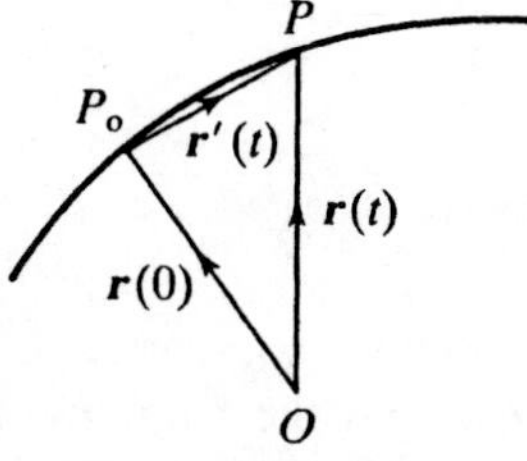

Fig 4.5 Position vector relative to fixed point of path.

When $t = 0$, $\mathbf{r}' = 0$ and so for $t \ll 1$ the function $\mathbf{r}'(t)$ can be approximated by a linear function

$$\mathbf{r}'(t) = \mathbf{p}t.$$

Generalizing the result on page 21, $\mathbf{p}$ is the derivative of $\mathbf{r}'$ evaluated at $t = 0$. Now $\mathbf{r}(0)$ is a constant vector and differentiating

$$\mathbf{r}'(t) = \mathbf{r}(t) - \mathbf{r}(0)$$

gives $\dot{\mathbf{r}}'(t) = \dot{\mathbf{r}}(t) = \mathbf{v}(t)$.

Hence $\mathbf{p} = \dot{\mathbf{r}}'(0) = \dot{\mathbf{r}}(0) = \mathbf{v}(0)$ so that

$$\mathbf{r}' = \mathbf{v}(0)t.$$

This equation represents a straight line in the direction of $\mathbf{v}(0)$, in fact the tangent to the path at the point P_0. The point P_0 was chosen arbitrarily and so we have shown that close to any point on the path of a given particle, the motion approximates to straight line motion, the direction of the velocity of the particle as it passes the point being in the direction of this straight line motion. This direction is called the **instantaneous direction of motion**, it is always tangential to the path of the particle. The distance of the particle P from P_0, in the linear approximation, is $|\overrightarrow{O'P}| = |\mathbf{r}'| = |\mathbf{v}(0)|t$. The rate of change of this distance with respect to time t is equal to $|\mathbf{v}(0)|$. It follows that the modulus of the velocity of the particle as it passes any point of the path is equal to the speed of the particle at that point. In writing down the above linear approximation a dimensionless time $\bar{t}$ should have been used; obtained, for example, by dividing t by the total time of travel of the particle.

Example I

The equation of the path of a particle P relative to an observer O is given by

$$\mathbf{r} = (ut, -2ut, \tfrac{1}{2}at^2)$$

where u and a are constants. Show that the minimum speed of the particle is $\sqrt{5}|u|$ and that the instantaneous directions of motion of the particle at times u/a and $-5u/a$ are perpendicular.

SOLUTION

The velocity of the particle is obtained by differentiating $\mathbf{r}$ with respect to time to give

$$\mathbf{v} = (u, -2u, at).$$

The speed of the particle at time t is

$$|\mathbf{v}| = \sqrt{5u^2 + a^2t^2}.$$

This is a minimum when $t = 0$ so that the minimum speed is $\sqrt{5}|u|$ as required. Notice the importance of the modulus; speed is always positive whereas the constant u could have either sign. The instantaneous direction of motion is the direction of the velocity. At time $t = u/a$,

$$\mathbf{v} = \mathbf{v}_1 = (u, -2u, u)$$

and at time $t = -5u/a$

$$\mathbf{v} = \mathbf{v}_2 = (u, -2u, -5u).$$

Using the scalar product,

$$\mathbf{v}_1.\mathbf{v}_2 = u^2 + 4u^2 - 5u^2 = 0$$

and so the two instantaneous directions of motion are perpendicular. •

Example 2

Two rods AB and BC, each of length a, are hinged at B and move in a plane with A fixed. Show that the squares of the speeds of B and C relative to A are $a^2\dot{\theta}^2$ and $a^2[\dot{\theta}^2 + \dot{\phi}^2 + 2\dot{\theta}\dot{\phi}\cos(\phi - \theta)]$ respectively, where θ and ϕ are the angles between a fixed line l lying in the plane and the rods AB and BC respectively, each angle being taken anticlockwise.

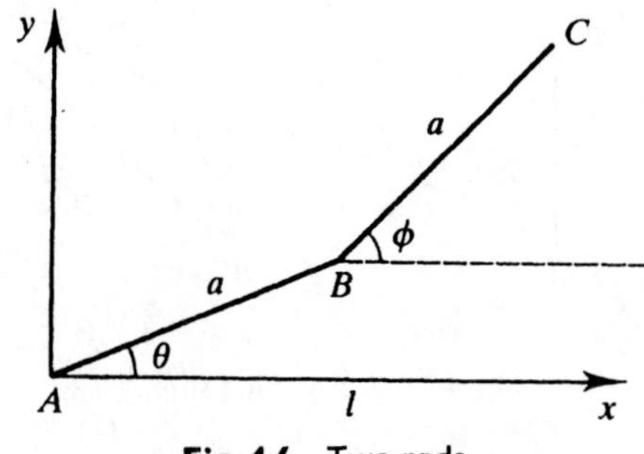

Fig 4.6 Two rods.

SOLUTION

Take cartesian axes with origin at A and x axis along l as in Fig 4.6. Then the coordinates of the points B and C are

$$(a\cos\theta, a\sin\theta) \quad \text{and} \quad (a\cos\theta + a\cos\phi, a\sin\theta + a\sin\phi),$$

respectively. Using the chain rule the velocities of the two points are

$$\mathbf{v}_B = (-a\sin\theta\,\dot{\theta}, a\cos\theta\,\dot{\theta})$$
$$\text{and} \quad \mathbf{v}_C = (-a\sin\theta\,\dot{\theta} - a\sin\phi\,\dot{\phi}, a\cos\theta\,\dot{\theta} + a\cos\phi\,\dot{\phi}).$$

The square of the speed is the square of the magnitude of the velocity, that is the sum of the squares of the components of the velocity. Hence

$$|\mathbf{v}_B|^2 = a^2\sin^2\theta\,\dot{\theta}^2 + a^2\cos^2\theta\,\dot{\theta}^2 = a^2\dot{\theta}^2$$
$$\begin{aligned}\text{and} \quad |\mathbf{v}_C|^2 &= (-a\sin\theta\,\dot{\theta} - a\sin\phi\,\dot{\phi})^2 + (a\cos\theta\,\dot{\theta} + a\cos\phi\,\dot{\phi})^2\\ &= a^2\dot{\theta}^2 + a^2\dot{\phi}^2 + 2a^2\dot{\theta}\dot{\phi}(\sin\theta\sin\phi + \cos\theta\cos\phi)\\ &= a^2[\dot{\theta}^2 + \dot{\phi}^2 + 2\dot{\theta}\dot{\phi}\cos(\phi - \theta)].\end{aligned}$$

•

Example 3

A ship is steaming on a bearing of $\theta°$ with speed u. On board the ship the wind is apparently blowing on a bearing of $\phi°$ with speed v. Find an expression for the actual wind speed in terms of u, v and $\phi - \theta$.

SOLUTION

A good strategy for tackling relative velocity problems is to first represent the given information in terms of directed line segments and then to use these directed line segments to form a vector triangle representing the addition law of relative velocities. The last stage is made almost trivial if a sensible labelling is used. For example here we shall label the ship, air and water as S, A and W respectively. Then the velocity of the ship relative to the water is represented in Fig 4.7(i). The length

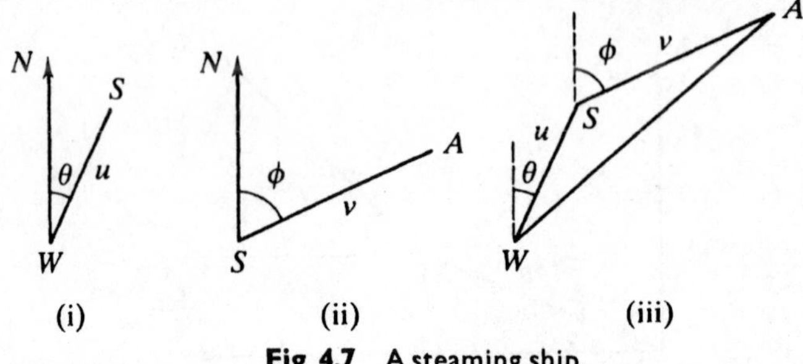

Fig 4.7 A steaming ship.

of the directed line segment is the magnitude of the velocity, that is the speed u. The velocity of the air relative to the ship is represented in Fig 4.7(ii). Using the two directed line segments to form a triangle gives Fig 4.7(iii). Notice that the labelling of the end points of the directed line segments makes it unnecessary to use an arrowhead to specify orientations. The third side of this triangle represents the velocity of the air relative to the water and the required actual wind speed is the length of this side. The angle $W\hat{S}A = \pi - (\phi - \theta)$ and so, using the cosine formula,

$$\text{actual wind speed} = \sqrt{u^2 + v^2 - 2uv\cos(\pi - (\phi - \theta))}$$

$$= \sqrt{u^2 + v^2 + 2uv\cos(\phi - \theta)}. \qquad \bullet$$

Example 4

A swimmer can swim a distance of 300m at a constant speed of 0.6ms^{-1} before becoming exhausted. What is the maximum current against which the swimmer could swim across a river 80m wide to a point directly opposite the point of entry into the water?

SOLUTION

Label the swimmer, water and river bank by S, W and B respectively. When swimming against the maximum current a distance of 300m will be covered at a speed of 0.6ms^{-1}. This will take 500s and in this time the swimmer must swim 80m relative to the river bank. The swimmer's speed relative to the river bank must therefore be 0.16ms^{-1}. This information is represented in Fig 4.8 together with the

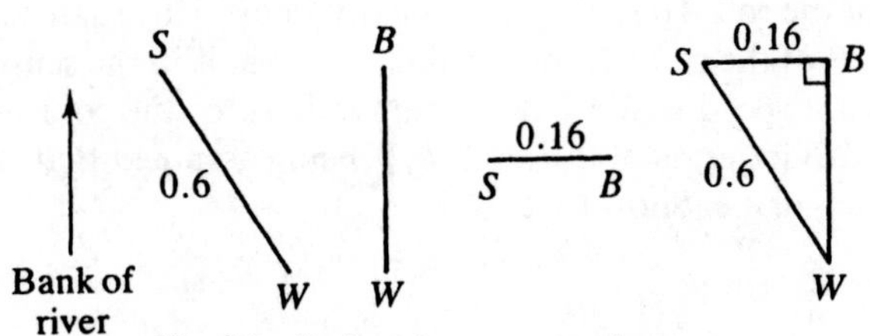

Fig 4.8 Swimming across a river.

vector triangle representing the addition law of relative velocities. The current is the magnitude of the velocity of the water relative to the bank and is therefore represented by the third side of the above triangle. By Pythagoras's theorem the maximum current is therefore

$$\sqrt{0.6^2 - 0.16^2} = 0.5\text{ms}^{-1}.$$

The derivative of the velocity $\mathbf{v}(t)$ is defined to be the **acceleration of P relative to O** and is written as $\dot{\mathbf{v}}$ or $\frac{d\mathbf{v}}{dt}$. This acceleration can also be written as a second derivative, namely $\ddot{\mathbf{r}}$ or $\frac{d^2\mathbf{r}}{dt^2}$. Assuming that the two observers O and O' are using identical clocks and frames of reference S and S' which are not rotating relative to each other, the addition law of relative velocities can be differentiated with respect to time to generalize the addition law of relative accelerations

$$\text{acc of } P \text{ rel to } O' = \text{acc of } P \text{ rel to } O + \text{acc of } O \text{ rel to } O',$$

obtained previously for motion on a straight line, to motion in space.

EXERCISES ON 4.1

1. Particles P and P' are located at points whose cartesian coordinates at time t are $(6t_0t, 5t^2, -2t_0t)$ and $(-7t_0t, 5t^2, 4t_0t)$ where t_0 is a constant. Write down expressions for the cartesian components of the velocities of P and P' relative to the origin O. Deduce that the velocity of P relative to O is at right angles to the position vector of P' relative to O at time $t = t_0$. Prove that P moves relative to P' on a straight line perpendicular to the y-axis with speed $\sqrt{205}t_0$.

2. A rod AB of length $2a$ is hinged at its mid point to one end of a rod CD of length a as shown in the following diagram.

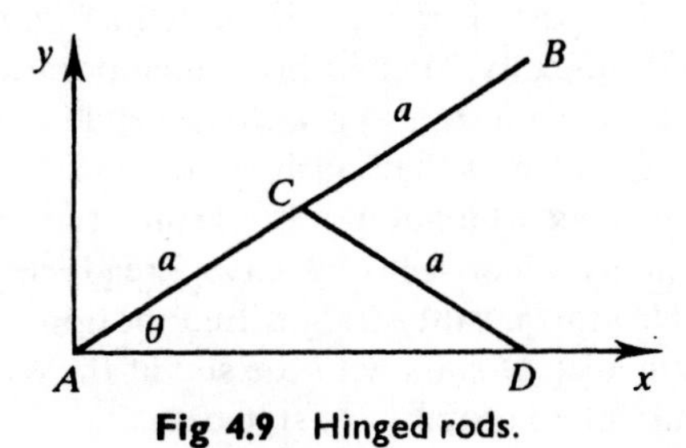

Fig 4.9 Hinged rods.

The end A of the rod AB is hinged at the origin of a cartesian coordinate system and the end D of the rod CD moves along the x-axis in the sense of increasing x with a constant speed v. Write down expressions for the coordinates of the end points D and B in terms of the angle θ, it being assumed that the motion takes place in the xy-plane. Show that

$$\dot{\theta} = -v/2a \sin \theta$$

and obtain expressions for the cartesian components of the velocity of B.

3. An aeroplane which can fly at a constant airspeed v flies a distance d due east against a wind blowing from the southeast with speed u. Show that the flight time is $\sqrt{2}d/(-u + \sqrt{2v^2 - u^2})$ and find the ratio of this time to the flight time for the return journey.

4. A ship is steaming due east at 10 knots and a flag on the mast head points south-east. The ship then turns south and the flag points east. What is the direction and speed of the wind?

5. A train moves on a straight horizontal line with a constant acceleration f. Prove that, measured relative to the train, the acceleration due to gravity acts at an angle $\tan^{-1} f/g$ with the true vertical and find an expression for the magnitude of this acceleration.

4.2 Inertial Frames and Newton's First Law of Motion

In section 1.2 our intuitive ideas about motion led us to the conclusion that if the resultant (nett) force acting on a particle is zero then the motion of the particle is uniform, that is the particle moves on a straight line with constant speed. A particle moving under a zero resultant force is said to be **moving freely** and the above conclusion is formulated as **Newton's first law of motion**:

- a freely moving particle moves on a straight line with constant speed

This law was first formulated by Galileo in 1638. We saw in the last section that straight line motion relative to one observer O will not appear to be straight line motion relative to another observer O' if O' is accelerating relative to O or if the frame of reference S' used by O' is rotating relative to the frame of reference S used by O. The statement of Newton's first law of motion given above is therefore incomplete, an observer O and frame of reference S needs to be identified relative to which the freely moving particle moves on a straight line with a constant speed. Newton recognized this difficulty. He and his contemporaries looked at the night sky and observed that the distant stars appear to be fixed in space. In order to make sense of the motion of the planets the Polish astronomer Copernicus had verified the postulate that the planets orbit about a fixed sun – the point of view which led to the persecution of many scientists as we have already seen. When referring to straight line motion Newton meant straight line motion relative to a frame of reference $S_{\odot}$ fixed to the distant stars with the sun at the origin. Hence Newton's first law of motion could more properly be stated as

- a freely moving particle moves on a straight line with constant speed relative to a frame of reference $S_\odot$ fixed to the distant stars with the sun at the origin

This form of the first law is well suited to celestial motions but is hardly practical for terrestrial motions, that is to motions of a particle moving close to the earth's surface. We know that the earth orbits about the sun and also rotates on its axis. This rotation may be neglected for terrestrial motions of short duration compared to the day and as we saw in the last section the motion of the earth in its orbit will approximate, using a linear approximation, to motion on a straight line with constant speed. It follows that for short duration terrestrial motions a frame $S_\oplus$ fixed to the earth will neither rotate nor accelerate relative to the frame $S_\odot$ and therefore for such motions Newton's first law of motion could be stated as

- a freely moving particle moves on a straight line with a constant speed relative to a frame of reference $S_\oplus$ fixed to the earth.

For terrestrial motions of longer duration the rotation of the earth must be taken into account, for example when discussing the motion of an intercontinental ballistic missile. For motions of even longer duration the motion of the earth on its orbit about the sun must also be taken into account, for example when calculating the splashdown of a space capsule.

The above approach picks out two particular frames of reference relative to which a freely moving particle moves uniformly. This is somewhat artificial and the approach is open to serious criticism because we now know that the distant stars are not fixed. They only appear fixed because they are so distant! This statement is in agreement with everyday experience of the theory of perspective. For example, if you view the countryside from a moving car then you will notice that the distant trees appear to move more slowly than the trees closer to the road. How could Newton's first law of motion be better stated? To answer this question we shall use the fact that a particle P will move on a straight line with constant speed relative to O if and only if its velocity relative to O is constant, that is if and only if its acceleration relative to O is zero. Now consider two observers O and O'. If their frames of reference S and S' are not rotating relative to each other then, according to the addition law of relative accelerations,

$$\mathbf{a}' = \mathbf{a} + \mathbf{A}$$

where $\mathbf{a}$ and $\mathbf{a}'$ are the accelerations of P relative to O and O' respectively, and $\mathbf{A}$ is the acceleration of O relative to O'. It follows from this addition law that if $\mathbf{a} = 0$ then $\mathbf{a}' = 0$ if and only if $\mathbf{A} = 0$. Hence if a particle is moving on a straight line with constant speed relative to the observer O then the particle is moving on a straight line with constant speed relative to the observer O' if and only if $\mathbf{A} = 0$, that is if and only if O moves on a straight line with constant speed relative to O'. It follows that if Newton's first law of motion holds relative to one observer it will hold relative to a whole infinite family of observers, each observer moving on a straight line with constant speed relative to the other observers, each frame of reference being non rotating relative to the other frames of reference. These observers are called **inertial observers** and their associated frames of reference are called **inertial frames**. Newton's first law of motion can now be stated as

- a freely moving particle moves on a straight line with constant speed relative to each inertial frame.

Here the phrase 'relative to each inertial frame' is taken to include the fact that the speed is relative to the origin of the frame. This usage of the phrase will be assumed throughout the book.

TUTORIAL PROBLEM 4.3

A particle is observed to be moving on a straight line with a constant speed relative to a given observer O and frame of reference S. Can you infer that the frame S is inertial and that the particle is moving freely? Discuss how Newton's first law of motion might be used to construct an inertial frame.

4.3 Newton's Second and Third Laws of Motion

When a force is acting on a particle the motion of that particle relative to each inertial frame deviates from the motion on a straight line with constant speed which the particle would experience if it were moving freely. In Section 2.2 we met **Newton's second law of motion** which can now be stated more precisely as

- the force acting on a particle is proportional to the product of the mass of the particle and the resulting acceleration of the particle relative to each inertial frame.

Strictly speaking the mass referred to here is the **inertial mass** of the particle, a refinement which will be discussed in Section 5.3. The **equation of motion** of a given particle P of mass m moving with acceleration $\mathbf{a}$ relative to a given inertial frame under the action of a force $\mathbf{F}$ expresses Newton's second law symbolically as

$$m\mathbf{a} = \mathbf{F}.$$

Here we have adopted the same conventions as in Section 2.2, placing the force on the right hand side of the equation and choosing the unit of force so that the constant of proportionality inherent in Newton's second law is equal to one. The equation of motion can also be written in either of the forms

$$m\ddot{\mathbf{r}} = \mathbf{F} \quad \text{or} \quad m\dot{\mathbf{v}} = \mathbf{F},$$

where $\mathbf{r}$ and $\mathbf{v}$ are the position vector and velocity of P relative to the given inertial frame.

TUTORIAL PROBLEM 4.4

Deduce from the equation of motion that if a particle is moving freely, so that $\mathbf{F} = 0$, then the particle will move on a straight line with constant speed relative to each inertial frame.

This tutorial problem shows that Newton's first law of motion is a consequence of the second law. It is important to realize that it is not redundant; it is the first law which enables us to identify the inertial frames of reference which must be used in the second law. In practice the inertial frame $S_{\odot}$ introduced in the last section will be used when discussing celestial motions and the inertial frame $S_{\oplus}$ will almost always be used when discussing terrestrial motions of short duration although for such motions other inertial frames sometimes prove to be more convenient.

In Section 2.2, fixed origins and observers at rest were introduced when discussing the mechanics of a particle moving on a straight line. These concepts were not actually defined but to be consistent with the general formulation of Newton's second law of motion we can now see that the origin must be fixed, and the observer must be at rest, relative to some inertial frame. Furthermore the straight line on which the particle moves must also be fixed relative to some inertial frame; if it is not, the discussion in Section 2.2 no longer applies.

Newton always wrote about bodies in motion and so might not have been fully aware that his laws apply to the motion of particles alone. He certainly did not write down the equation of motion in any of the forms given above because vectors were not introduced until much later. Neither did he write down the scalar equations

$$m\ddot{x} = F_x, m\ddot{y} = F_y, m\ddot{z} = F_z$$

which are the three cartesian components of the vector equation of motion. It was the Swiss mathematician Leonard Euler (1707–1783) who presented these three equations in 1752 and who later referred to them as the 'first principles of mechanics'. Applying these equations to the elements of mass making up a rigid body enabled Euler to extend the laws of mechanics to cover the motion of bodies of finite size.

Newton's second law of motion involves the force acting on the particle and presupposes that forces can be modelled as vectors. In most cases several different forces will be acting on a given particle simultaneously and it is necessary to postulate that the force referred to in Newton's second law is the vector sum of all the individual forces; this is the so called **resultant force** acting on the particle. The necessity for making such a postulate was first realized by Ernst Mach (1838–1916) in 1883. The name of this Austrian physicist and psychologist may be familiar to you from the mach number associated with supersonic flight.

The determination of the actual forces acting on a given particle is the subject of the other branches of science. Newton himself gave the expression for the gravitational attraction between two particles, this will be discussed in the next Chapter. The other insight which Newton gave into the nature of force is incorporated in his **third law of motion,**

- the force exerted on one particle by another is equal and opposite to the force exerted by the first particle on the second.

Here the phrase equal and opposite is an abbreviation for equal in magnitude and opposite in direction. This law is often quoted in the form

- action and reaction are equal and opposite.

In fact, not all forces acting between particles satisfy Newton's third law. For example, the force which a segment of one electrical circuit exerts on a segment of a second electrical circuit is not equal and opposite to the force that the segment of the second electrical circuit exerts on the segment of the first. Nevertheless most forces between particles do obey the third law and such forces are sometimes called **Newtonian.**

Newton's third law of motion as stated above is the **weak form** of the third law. There is a **strong form** of the third law which is stated as

- the force exerted on one particle by another is equal and opposite to the force exerted by the first particle on the second and acts along the line joining the particles.

Some readers might be somewhat puzzled by the fact that Newton's second law alone was introduced in Chapter 1 when discussing straight line motion. For such motion the impact of the first law is somewhat reduced because the particle moves on a straight line whether or not it is moving freely; all that then remains of the first law is that a freely moving particle moves with constant speed. The third law was implicit in some of the earlier chapters, for example the external force exerted on one end of an elastic string and the internal force, that is the tension, exerted on that end are equal and opposite. The full significance of the third law becomes apparent when discussing the motion of a system of particles.

Isaac Newton was born on Christmas Day, 1642. The manor house in which he was born still stands in the village of Woolsthorpe, 7 miles from Grantham in Lincolnshire. It is open to the public and administered by the National Trust. His life spanned the reigns of six monarchs and also the Commonwealth under the Protectors Oliver Cromwell and his son Richard. That Newton maintained such a high profile as the Lucasian Professor of Mathematics at Cambridge, a Member of Parliament, Master of the Royal Mint and President of the Royal Society during one of the most stormy periods of English history speaks much for his strength of character and for the esteem in which he was held both at home and abroad. Many of his most notable contributions to science and mathematics were formulated during the plague years when the University of Cambridge was closed because of the fear of contagion. Newton continued his studies at Woolsthorpe during those years and 1666 is often referred to as his *annus mirabilis* (year of miracles). During that year he discovered and developed the binomial series, the differential calculus, the theory of colours, the integral calculus and the universal theory of gravitation. This was an astounding series of achievments all completed before his 24th birthday! Equally astounding is the fact that the wills of his father, grandfather and uncles still exist and are all signed with a cross; Isaac Newton was the first male member of his family to be literate! Despite the richness of his scientific work he actually wrote more on church history and chronology than on any other subject and he was also deeply interested in alchemy, an interest which was largely hidden by his early biographers.

Newton was persuaded by his pupil Edmund Halley – of comet fame – to publish his work on mechanics as a three volume treatise known as the Principia. The first edition appeared in 1687 and revolutionized the understanding of motion, introducing the laws of mechanics and their applications to, for example, motion in a resisting medium and celestial mechanics. Despite having invented the calculus, Newton used geometrical arguments in the Principia and this may have disadvantaged British scientists in so far as further developments of mechanics were concerned. Suffice it to say that most of the fundamental developments made in the eighteenth century were due to Continental mathematicians using Leibnitz's notation for the calculus rather than the more cumbersome notation introduced by Newton. Leonard Euler was one of the most notable of the earlier of these Continental mathematicians. In his earlier days Euler was a Professor of Geometry at St. Petersberg and he not only extended Newton's laws of motion to bodies of finite size but also gave new and deep insights into many other branches of mechanics, for example the vibration of a string and the motion of a fluid. In 1741 he moved to Berlin to work under the patronage of Frederick the Great, King of Prussia. Royal patronage carried obligations and one of Euler's duties was to instruct the Princess d'Anhalt Deffau, a niece of Frederick the Great, by letter. These letters were later published under the title 'Letters to a German Princess' and in them Euler gives a succinct account of mechanics, astronomy, optics, the theory of sound and many other topics. In the letter of 17th February, 1761 he uses Venn diagrams some hundred years before the birth of John Venn in the city of Kingston upon Hull.

Euler, like Galileo, lost his sight in later life but this did nothing to diminish his scientific output. He worked in the midst of his family, helped in his blindness by his sons and pupils. This contrasts sadly with the later years of Newton's bachelor life which were often soured by bitter quarrels and disputes with fellow mathematicians. One such quarrel was with Robert Hooke, the curator of experiments to the Royal Society. Newton refused to accept the Presidency of the Royal Society until after Hooke's death and one of his first acts as President was to eliminate any tangible evidence of Hooke's long involvement with the Society by destroying the apparatus used by Hooke in his public demonstrations.

4.4 Integration of the Equation of Motion

Given the resultant force $\mathbf{F}$ acting on a given particle P, the equation of motion

$$m\ddot{\mathbf{r}} = \mathbf{F}$$

becomes a second order differential equation for the position vector $\mathbf{r}$ of P. This equation can be integrated directly using vectorial techniques and its general solution will contain two arbitrary constant vectors of integration. Alternatively, the three cartesian components

$$m\ddot{x} = F_x, m\ddot{y} = F_y, m\ddot{z} = F_z$$

of the equation can be integrated. Each is a second order differential equation whose general solution will involve two arbitrary constants of integration. The resulting six arbitrary constants of integration are either equal or related to the cartesian components of the two arbitrary constant vectors of integration which arise when vectorial techniques are used to integrate the equation of motion. The general solution of the equation of motion describes all possible motions of the particle consistent with the given force $\mathbf{F}$ (a motion is determined by a function $\mathbf{r}(t)$). To determine the actual motion of a given particle, initial conditions are required in order to calculate explicit values of the two constant vectors of integration.

In general the force $\mathbf{F}$ acting on the particle P will be a function of the position vector $\mathbf{r}$, velocity $\mathbf{v}$ and time t, so that

$$\mathbf{F} = \mathbf{F}(t, \mathbf{r}, \dot{\mathbf{r}}).$$

Given this function the resulting equation of motion can be integrated using the numerical techniques introduced in Section 2.2, either directly or applied to each cartesian component of the equation. This will not be carried through here but an important consequence, analogous to that in Section 2.2, is that the values $\mathbf{r}_0$ and $\mathbf{v}_0$ of the position vector and velocity of the particle at some given instant of time t_0 are suitable initial conditions in terms of which the constant vectors of integration can be determined. As we saw in the case of straight line motion these are not the only initial conditions, others are equally suitable for the determination of the constant vectors of integration.

Example 5

A projectile P is fired from a point O on the earth's surface with speed u at an angle of θ radians with the horizontal. Find an expression for the position vector $\mathbf{r}$ of P relative to O as a function of time t, air resistance being neglected.

SOLUTION

Neglecting air resistance the only force acting on the projectile is its weight which has constant magnitude mg and a fixed direction. This force can be written vectorially as $m\mathbf{g}$ where the constant vector $\mathbf{g}$ is in the direction of the downward vertical. Taking O as the origin of a frame of reference $S_\oplus$ fixed to the earth's surface, the equation of motion of the projectile can be written as

$$m\ddot{\mathbf{r}} = m\mathbf{g}$$

or, cancelling m,

$$\ddot{\mathbf{r}} = \mathbf{g}.$$

Repeated integration of this equation yields the following expressions for the velocity and position vector of the projectile relative to O at time t,

$$\dot{\mathbf{r}} = \mathbf{g}t + \mathbf{c}_1$$

$$\text{and} \qquad \mathbf{r} = \tfrac{1}{2}\mathbf{g}t^2 + \mathbf{c}_1 t + \mathbf{c}_2.$$

Here $\mathbf{c}_1$ and $\mathbf{c}_2$ are constant vectors of integration. Assuming that the projectile is projected from the point O at time $t = 0$ it follows from the first expression that $\mathbf{c}_1$ is the velocity of projection $\mathbf{u}$ and from the second expression that $\mathbf{c}_2$ is zero giving

$$\mathbf{r} = \tfrac{1}{2}\mathbf{g}t^2 + \mathbf{u}t.$$

Comparing this equation with the vector equation of a plane passing through the origin and containing the vectors **a** and **b**, namely

$$\mathbf{r} = \lambda\mathbf{a} + \mu\mathbf{b},$$

where λ and μ are two parameters, shows that the path of the projectile lies in the plane passing through the point of projection and containing the vectors **g** and **u**. Since **g** is in the direction of the downward vertical this plane of motion is a vertical plane. Now introduce cartesian axes Ox and Oy in the plane of motion with Ox horizontal and Oy vertical and Oz perpendicular to the plane of motion as in Fig. 4.10.

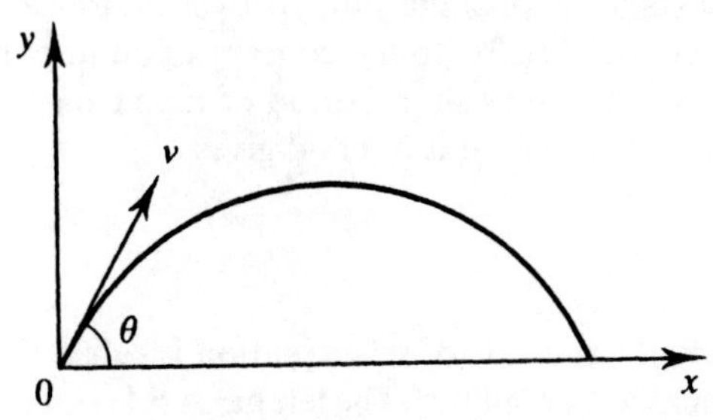

Fig 4.10 Path of projectile.

Then $\mathbf{r} = (x, y, 0), \mathbf{g} = (0, -g, 0)$ and $\mathbf{u} = (u \cos \theta, u \sin \theta, 0)$ so that the three cartesian components of the equation

$$\mathbf{r} = \tfrac{1}{2}\mathbf{g}t^2 + \mathbf{u}t$$

are
$$x = u \cos \theta \; t$$
$$y = -\tfrac{1}{2}gt^2 + u \sin \theta \; t$$
and
$$z = 0.$$

The second of these equations is essentially the equation obtained previously for the displacement of a projectile moving vertically under gravity; note that the displacement was then denoted by x rather than by y. Indeed, relative to a second inertial observer O', moving along the straight line Ox with constant speed $u \sin \theta$, the projectile would appear to be moving vertically. •

Integration of the equation of motion using vectorial techniques is rarely as simple as in the above example and is best learnt in the context of specific situations.

Unless otherwise stated the mass of a given particle will be taken to be constant so that the equation of motion of a particle of mass m moving with velocity **v** relative to an inertial observer can be rewritten in the form

$$\frac{d}{dt}(m\mathbf{v}) = \mathbf{F}.$$

The vector $m\mathbf{v}$ is defined to be the **linear momentum** of the particle and will be denoted by $\mathbf{p}$ so that

$$\frac{d\mathbf{p}}{dt} = \mathbf{F}.$$

Newton himself expressed his second law of motion in terms of linear momentum – what he called quantity of motion – rather than acceleration. This alternative expression can be stated in the form

- the rate of change of the linear momentum of a particle relative to each inertial frame is proportional to the force acting on the particle.

You will recall the discussion in Section 1.2 of our intuitive ideas about the concept of force and the importance of distinguishing between impulsive and continuously acting forces. Let us return, briefly, to the concept of an impulsive force, that is a force $\mathbf{F}$ which acts only for a very short period of time from, say, $t = t_1$ to $t = t_2$. Integrating the last equation over this interval gives

$$\int_{t_1}^{t_2} \frac{d\mathbf{p}}{dt}\,dt = \int_{t_1}^{t_2} \mathbf{F}\,dt.$$

The integral on the right hand side of this equation is called the **impulse** acting on the particle, and is usually denoted by $\mathbf{I}$. The left hand side can be integrated to yield $[\mathbf{p}]_{t_1}^{t_2}$ so that the impulse acting on the particle is equal to the change in the linear momentum of the particle. This result can be written in terms of the velocity of the particle as

$$\mathbf{I} = m\mathbf{v}_2 - m\mathbf{v}_1$$

where $\mathbf{v}_1$ and $\mathbf{v}_2$ are the velocities of the particle immediately before and after the impulse has acted on the particle. If the impulse acts on a particle which is initially at rest then $\mathbf{v}_1 = 0$ and we see that doubling the impulsive force $\mathbf{F}$ will double the impulse and therefore double the velocity $\mathbf{v}_2$ with which the particle starts to move, as we saw intuitively in Section 1.2.

Let us now return to a particle moving under a continuously acting force $\mathbf{F}$ and the equation of motion in the form

$$\frac{d\mathbf{p}}{dt} = \mathbf{F}.$$

Suppose that the resultant force $\mathbf{F}$ has a zero component in a given fixed direction, that is in a direction fixed relative to the inertial frame being used. If $\hat{\epsilon}$ is a unit vector in this direction, then

$$\mathbf{F}.\hat{\epsilon} = 0$$

and taking the scalar product of the equation of motion with $\hat{\epsilon}$ yields

$$\frac{d\mathbf{p}}{dt}.\hat{\epsilon} = 0.$$

Now $\hat{\epsilon}$ is a constant vector and so can be taken inside the derivative; formally you need to use the product rule for the derivative of a scalar product in order to carry out this procedure. The resulting equation

$$\frac{d}{dt}(\mathbf{p}.\hat{\epsilon}) = 0$$

can be integrated trivially to yield

$$\mathbf{p}.\hat{\epsilon} = \text{constant}$$

so that if the resultant force has a zero component in a given fixed direction then the corresponding component of the linear momentum remains constant throughout the motion of the particle. This result is known as the **conservation of the component of linear momentum.** If **F** vanishes then the linear momentum as a whole is conserved.

Taking the vector product of the equation of motion with **r** gives

$$\mathbf{r} \times \frac{d\mathbf{p}}{dt} = \mathbf{r} \times \mathbf{F}.$$

The vector $\mathbf{r} \times \mathbf{F}$ is called the **moment of the force F about the origin** and will be denoted by **M**. Using the product rule

$$\frac{d}{dt}(\mathbf{r} \times \mathbf{p}) = \mathbf{r} \times \frac{d\mathbf{p}}{dt} + \frac{d\mathbf{r}}{dt} \times \mathbf{p}$$

and noting that

$$\frac{d\mathbf{r}}{dt} \times \mathbf{p} = \mathbf{v} \times (m\mathbf{v}) = m\mathbf{v} \times \mathbf{v} = 0,$$

it follows that

$$\frac{d}{dt}(\mathbf{r} \times \mathbf{p}) = \mathbf{r} \times \mathbf{F}.$$

The vector $\mathbf{r} \times \mathbf{p}$ is the moment of the linear momentum about the origin. It is called the **angular momentum of the particle about O** and is denoted by **L**. Hence

$$\frac{d\mathbf{L}}{dt} = \mathbf{M}$$

so that

- the rate of change of the angular momentum of a particle about a given origin is equal to the moment of the force acting upon the particle about the same origin.

If the moment of the resulting force has a zero component in a given fixed direction then the corresponding component of the angular momentum remains constant throughout the motion of the particle. This result is known as the **conservation of the component of the angular momentum.** If **M** vanishes then the angular momentum as a whole is conserved.

TUTORIAL PROBLEM 4.5

A force **F** acts at the point P with position vector **r** relative to a given origin O. The line of action of the force is defined to be the straight line passing through P in the direction of $\mathbf{F}$, i.e. it is the line along which you would normally draw the directed line segment representing the force. Show that the magnitude of the moment of the force about O is equal to the magnitude $|\mathbf{F}|$ of the force times the perpendicular distance of O from the line of action of $\mathbf{F}$. Deduce that the moment of the force about O is given by $\mathbf{r}' \times \mathbf{F}$, where $\mathbf{r}'$ is the position vector of any point on the line of action of $\mathbf{F}$.

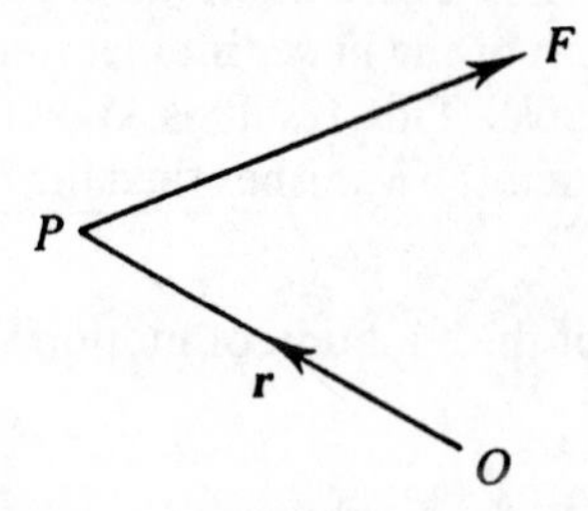

Fig 4.11 Line of action of a force.

If you have previously met the two dimensional concept of the moment of a force, discuss how the sign of the two dimensional moment is related to the direction of $\mathbf{M}$. Otherwise discuss how the concept of a moment is related to the balance equation governing the equilibrium of a balance.

Example 6

The cartesian coordinates (x, y, z), relative to an inertial frame, of a moving particle of mass m are given at time t by the equations

$$x = a \sin \omega t, \quad y = a \cos \omega t, \quad z = 2a\omega t,$$

where a and ω are constants. Write down expressions for the components of the linear momentum of the particle, the force acting on the particle, the moment of the force about the origin and the angular momentum of the particle about the origin. Verify that the moment of the force about the origin is equal to the rate of change of the angular momentum of the particle about the origin.

SOLUTION

The position vector $\mathbf{r}$ of the particle is given by

$$\mathbf{r} = (a \sin \omega t, a \cos \omega t, 2a\omega t).$$

The linear momentum of the particle is therefore

$$\mathbf{p} = m\dot{\mathbf{r}} = m(a\omega \cos \omega t, -a\omega \sin \omega t, 2a\omega).$$

The force acting on the particle is given by

$$\mathbf{F} = m\ddot{\mathbf{r}} = m(-a\omega^2 \sin \omega t, -a\omega^2 \cos \omega t, 0)$$

and the moment of this force about the origin is given by

$$\mathbf{M} = \mathbf{r} \times \mathbf{F} = m(2a^2\omega^3 t \cos \omega t, -2a^2\omega^3 t \sin \omega t, 0).$$

The angular momentum of the particle about the origin is defined as $\mathbf{L} = \mathbf{r} \times \mathbf{p}$ and so

$$\mathbf{L} = m(2a^2\omega \cos \omega t + 2a^2\omega^2 t \sin \omega t, 2a^2\omega^2 t \cos \omega t - 2a^2\omega \sin \omega t, -a\omega^2).$$

By inspection $\dot{\mathbf{L}} = \mathbf{M}$, as required. Notice also that both $\mathbf{F}$ and $\mathbf{M}$ have zero z-components and that the z-components of $\mathbf{p}$ and $\mathbf{L}$ are each constant as required by the appropriate conservation laws. •

Example 7

A particle P is moving with a constant angular momentum $\mathbf{L}$ about the origin. Prove that the path of P lies on a plane containing O and that as the particle moves the rate of change of the area swept out by the line segment OP is a constant.

SOLUTION
Taking the scalar product of $\mathbf{L} = \mathbf{r} \times \mathbf{p}$ with $\mathbf{r}$ yields

$$\mathbf{r}.\mathbf{L} = \mathbf{r}.(\mathbf{r} \times \mathbf{p}) = 0,$$

since any scalar triple product with a repeated factor is zero. The vector $\mathbf{L}$ is a constant and so comparing the above with the standard equation of a plane, i.e.

$$\mathbf{r}.\mathbf{n} = p,$$

shows that the path of the particle lies on a plane containing O and perpendicular to $\mathbf{L}$. Consider the area δA swept out by OP in a small time interval δt. From Fig 4.12 it follows that

$$\delta A = |\tfrac{1}{2}\mathbf{r} \times \delta\mathbf{r}|$$

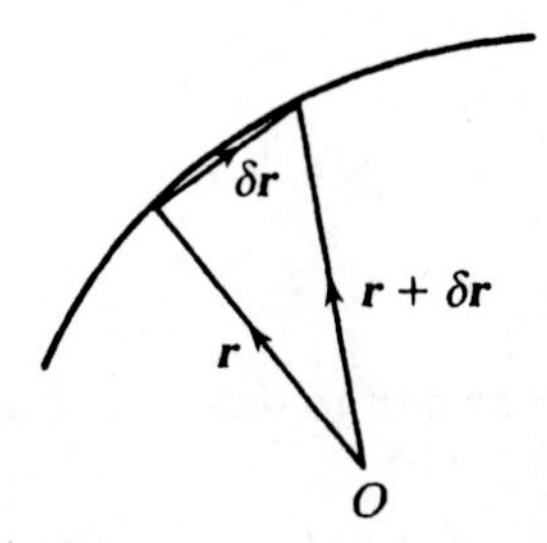

Fig 4.12 Area swept out by OP.

so that

$$\frac{dA}{dt} = \lim_{\delta t \to 0} \frac{\delta A}{\delta t} = \lim_{\delta t \to 0} |\tfrac{1}{2}\mathbf{r} \times \frac{\delta \mathbf{r}}{\delta t}|$$

$$= \tfrac{1}{2}|\mathbf{r} \times \dot{\mathbf{r}}| = \frac{|\mathbf{L}|}{2m}.$$

Since both $|\mathbf{L}|$ and m are constants it follows that dA/dt is also constant, as required. •

TUTORIAL PROBLEM 4.6

Deduce from the last example that the line segment OP sweeps out equal areas in equal times.

Conservation laws are very important because they enable us to write down first integrals of the equation of motion without having to carry out any formal integration. Another conservation law was met in Section 2.4, that is the conservation of the total energy of a particle moving on a straight line under the action of a force $F(x)$. To investigate the existence of an analogous conservation law for motion in space, suppose that the force $\mathbf{F}$ acting on a given particle is a function of the position vector of the particle so that the equation of motion takes the form

$$m\ddot{\mathbf{r}} = \mathbf{F}(\mathbf{r}).$$

The procedure used in Section 2.4 generalizes to three dimensions. This procedure involves multiplying the equation of motion by the velocity and integrating with respect to time. Of course there are two possible multiplications corresponding to the scalar product and the vector product. Since the energy of a particle is a scalar quantity it is the scalar product which is relevant here. The procedure therefore leads to the equation

$$\int m\ddot{\mathbf{r}}.\dot{\mathbf{r}}\,dt = \int \mathbf{F}.\dot{\mathbf{r}}\,dt.$$

The left hand side of this equation can be integrated, although not quite so trivially as in Section 2.4. In terms of the velocity $\mathbf{v}$ of the particle

$$\int m\ddot{\mathbf{r}}.\dot{\mathbf{r}}\,dt = \int m\frac{d\mathbf{v}}{dt}.\mathbf{v}\,dt.$$

For any vector $\mathbf{a}$,

$$\mathbf{a}.\mathbf{a} = |a|^2.$$

Differentiating this equation with respect to time, using the product rule, gives

$$\mathbf{a}.\frac{d\mathbf{a}}{dt} + \frac{d\mathbf{a}}{dt}.\mathbf{a} = 2|\mathbf{a}|\frac{d|\mathbf{a}|}{dt}.$$

Since the scalar product is commutative it follows that

$$\mathbf{a}.\frac{d\mathbf{a}}{dt} = |\mathbf{a}|\frac{d|\mathbf{a}|}{dt}.$$

This result often proves useful, it holds for all vectors $\mathbf{a}$. With $\mathbf{a}$ replaced by the velocity $\mathbf{v}$ and $|\mathbf{a}|$ by the speed $|\mathbf{v}|$ we can use the result to give

$$\int m\ddot{\mathbf{r}}.\dot{\mathbf{r}}\,dt = \int m|\mathbf{v}|\frac{d|\mathbf{v}|}{dt}\,dt$$

$$= \int m|\mathbf{v}|\,d|\mathbf{v}| = \tfrac{1}{2}m|\mathbf{v}|^2 + \text{constant}.$$

The expression $\frac{1}{2}m|\mathbf{v}|^2$ is familiar from Section 2.4, it is the kinetic energy of the particle. It follows that

$$\tfrac{1}{2}m|\mathbf{v}|^2 = \int \mathbf{F}.\dot{\mathbf{r}}\,dt = \int \mathbf{F}.\frac{d\mathbf{r}}{dt}dt = \int \mathbf{F}.d\mathbf{r}.$$

In order to integrate, in Section 2.4, an integral analogous to that on the right hand side of this equation, namely

$$\int F\,dx,$$

it was found to be convenient to define a function $V(x)$ by the equation

$$F(x) = -\frac{dV}{dx}.$$

An analogous function in space would be of the form $V(\mathbf{r})$. Such a function is called a scalar field and the natural derivative of a scalar field is the gradient which we denote by ∇V and which is defined to be the vector field with cartesian components

$$\left(\frac{\partial V}{\partial x}, \frac{\partial V}{\partial y}, \frac{\partial V}{\partial z}\right).$$

The equation defining $V(\mathbf{r})$, analogous to the above equation for motion on a straight line, is therefore

$$\mathbf{F}(\mathbf{r}) = -\nabla V.$$

The analogy with Section 2.4 now breaks down because this equation, unlike its one dimensional counterpart does not always admit a solution for $V(\mathbf{r})$. If a solution exists the force $\mathbf{F}(\mathbf{r})$ is defined to be **conservative** and the function $V(\mathbf{r})$ is then called the **potential energy** of the particle or the **potential** of the force. Understanding the conditions under which a given force $\mathbf{F}(\mathbf{r})$ is conservative requires results from vector analysis which you have not met yet. These conditions will not be discussed here. Nevertheless we know that if the force $\mathbf{F}(\mathbf{r})$ is conservative then

$$\begin{aligned}
\tfrac{1}{2}m|\mathbf{v}|^2 &= -\int \nabla V.d\mathbf{r} \\
&= -\int \left(\frac{\partial V}{\partial x}, \frac{\partial V}{\partial y}, \frac{\partial V}{\partial z}\right).(dx, dy, z) \\
&= -\int \left(\frac{\partial V}{dx}dx + \frac{\partial V}{\partial y}dy + \frac{\partial V}{\partial z}dz\right) \\
&= -\int dV \\
&= -V(\mathbf{r}) + \text{constant}.
\end{aligned}$$

It follows that the total energy E of the particle, defined by

$$\tfrac{1}{2}m|\mathbf{v}|^2 + V(\mathbf{r}) = E,$$

is a constant. This equation expresses the conservation of the total energy of the particle; it is usually referred to as the **energy equation.**

Example 8

Prove that for a force of the form

$$\mathbf{F}(\mathbf{r}) = F(r)\hat{\mathbf{r}},$$

the energy equation takes the form

$$\tfrac{1}{2}m|\mathbf{v}|^2 - \int F(r)dr = E.$$

Evaluate the potential energy $V(\mathbf{r})$ in the cases

$$F(r) = r \quad \text{and} \quad F(r) = r^{-2}.$$

SOLUTION

Taking the scalar product of the equation of motion

$$m\ddot{\mathbf{r}} = F(r)\hat{\mathbf{r}}$$

with $\dot{\mathbf{r}}$ and integrating with respect to time as in the text yields

$$\tfrac{1}{2}m|\mathbf{v}|^2 = \int F(r)\hat{\mathbf{r}}.\frac{d\mathbf{r}}{dt}dt.$$

We saw that for any vector $\mathbf{a}$

$$\mathbf{a}.\frac{d\mathbf{a}}{dt} = |\mathbf{a}|\frac{d|\mathbf{a}|}{dt}.$$

The magnitude of the position vector $\mathbf{r}$ of a point P relative to O is just the distance r of the point P from O so that

$$\mathbf{r}.\frac{d\mathbf{r}}{dt} = r\frac{dr}{dt}.$$

Hence

$$\int F(r)\hat{\mathbf{r}}.\frac{d\mathbf{r}}{dt}dt = \int \frac{F(r)}{r}\mathbf{r}.\frac{d\mathbf{r}}{dt}dt$$

$$= \int \frac{F(r)}{r}r\frac{dr}{dt}dt = \int F(r)dr$$

so that the energy equation takes the form

$$\tfrac{1}{2}m|\mathbf{v}|^2 - \int F(r)dr = E.$$

If $F(r) = r$ then $\int F(r)dr = \int r\,dr = \frac{1}{2}r^2 + \text{constant}$.

If $F(r) = r^{-2}$ then $\int F(r)dr = \int r^{-2}dr = -r^{-1} + \text{constant}$.

It follows that the required potential energies are given by

$$V(\mathbf{r}) = -\tfrac{1}{2}r^2 + \text{constant}$$

and $V(\mathbf{r}) = r^{-1} + \text{constant}.$

If you know how to calculate the gradient of a vector field you should verify that in each case $\mathbf{F}(\mathbf{r}) = -\nabla V$.

EXERCISES ON 4.4

1. Using the expressions obtained in Example 5 for the cartesian coordinates at time t of a projectile P fired from a point O on the earth's surface with speed u at an angle of θ radians to the horizontal, prove that

   (i) the greatest height of the projectile is $u^2 \sin^2 \theta/2g$

   (ii) the total time of flight of the projectile is $2u \sin \theta/g$

   (iii) the range of the projectile, i.e. total horizontal distance travelled, is $2u^2 \sin \theta \cos \theta/g$.

2. Show that if the projectile in the previous question is fired with the same speed and elevation but from a height H then the range of the projectile is increased by an amount

$$\frac{u \cos \theta}{g}\left(\sqrt{u^2 \sin^2 \theta + 2Hg} - u \sin \theta\right).$$

3. What is the SI unit of impulse?

4. Show that the angular momentum $\mathbf{L}$ of a particle P about an origin O, fixed in some inertial frame, is constant if and only if the force acting on P is directed towards or away from O.

5. The cartesian coordinates (x, y, z) of a moving particle, relative to an inertial frame, are given at time t by the equations

$$x = at, \quad y = bt^2, \quad z = ct^3,$$

   where a, b and c are constants. The mass of the particle is m. Find expressions for the force acting on the particle, the moment of this force about the origin and the angular momentum of the particle about the origin. Verify that the moment of the force is equal to the rate of change of the angular momentum.

6. Repeat Example 6 of Chapter 2 but without the assumption that B is pulled in the direction of the string. Put $\overrightarrow{AB} = \mathbf{r}$ and give your answer in terms of λ, r and l_0.

Summary

- the position of a particle P relative to an origin or observer O is specified by the **position vector** $\mathbf{r} = \overrightarrow{OP}$
- the **velocity** of P relative to O is the derivative of the position vector
- the **acceleration** of P relative to O is the derivative of the velocity
- the magnitude of the velocity is **speed** and the direction of velocity is the **instantaneous direction of motion**, tangential to the path of the particle

- in order to differentiate a vector function of time it is necessary to introduce a **frame of reference**
- the addition laws of relative velocities and of relative accelerations take the same form as for straight line motion provided that the frames of reference associated with the two observers are not rotating relative to each other
- a freely moving particle moves on a straight line with constant speed relative to each inertial frame; this is **Newton's first law**
- **inertial frames of reference** do not rotate relative to each other and the origin of each moves on a straight line with constant speed relative to the others
- when discussing celestial motions the frame of reference with the sun as origin and axes fixed relative to the distant stars is taken to be inertial
- when discussing terrestrial motions of small duration compared to the day any frame of reference fixed relative to the earth is taken to be inertial
- the **equation of motion** $m\ddot{\mathbf{r}} = \mathbf{F}$ expresses **Newton's second law,** namely the force acting on a particle is proportional to the mass of the particle and to the resulting acceleration of the particle relative to each inertial frame.
- the force **F** appearing in the equation of motion is the vector sum of the individual forces acting on the particle; this sum is the **resultant force**
- the force exerted on one particle by another is equal and opposite to the force exerted by the first particle on the second; this is **Newton's third law** in its **weak form**
- the **strong form** of Newton's third law is obtained by adding that the force between the particles acts along the line joining the particles
- $\mathbf{p} = m\mathbf{v}$ is the **linear momentum**
- $\mathbf{L} = \mathbf{r} \times \mathbf{p}$ is the **angular momentum** about O
- $\mathbf{M} = \mathbf{r} \times \mathbf{F}$ is the moment of the force **F** about O
- if the force/moment has zero component in a fixed direction then the corresponding component of the linear /angular momentum is conserved
- a force **F** is said to be **conservative** if it can be written in the form $\mathbf{F} = -\nabla V$; $V(\mathbf{r})$ is the **potential** of the force or **potential energy** of the particle
- for a conservative force the energy equation

$$\tfrac{1}{2}m|v|^2 + V(\mathbf{r}) = E$$

expresses the conservation of energy.

FURTHER EXERCISES

1. Particles P and P' are located at points whose cartesian coordinates at time t are $(0, u_0t, -\frac{1}{2}gt^2)$ and $(v_0t, 0, -\frac{1}{2}gt^2)$, where u_0, v_0 and g are constants. Write down expressions for the cartesian components of the velocities of P and P' and hence

deduce that at time $t = 0$ the paths of the particles are at right angles to each other but that subsequently the paths will become almost parallel. Prove that P moves relative to P' in a straight line with speed $\sqrt{u_0^2 + v_0^2}$.

2. A particle is observed to be moving in the x, y plane radially away from the origin O with a constant speed v. Find the cartesian components of the velocity of the particle as it passes the point with cartesian coordinates $(4a, 3a)$.

3. A straight rod AB of length $2a$ is placed over a wall of height $\frac{1}{2}a$, as shown in Fig 4.13. The end A approaches the foot of the wall, along a horizontal floor, with a speed v. Prove that the rate of change of the angle θ with respect to time is given by

$$a\dot{\theta} = 2v \sin^2 \theta,$$

motion being assumed to take place in a vertical plane.

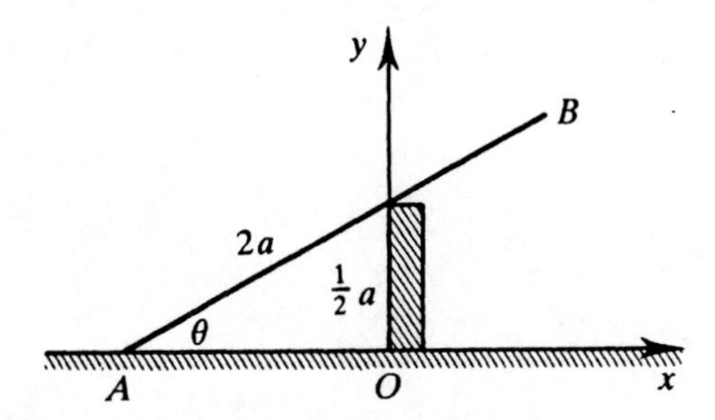

Fig 4.13 Rod resting on wall.

Obtain expressions for the cartesian coordinates x, y of the end B (see diagram) in terms of the angle θ and hence find the velocity of B. Show that when $\theta = 30°$ the speed of B is equal to the speed of A.

4. A jogger always runs with a constant speed of 3ms^{-1}. Whilst running along a straight road the jogger feels the wind to be blowing from left to right. After turning through an angle of 90° the jogger feels the wind to be blowing from right to left. What is the wind speed?

5. A bird is sitting on the top of a flag-pole of height 10m. A stone is thrown at the bird at an angle of projection of 60° to the horizontal from a point at ground level 6m from the foot of the flag-pole. At the instant of projection, the bird flies off with a constant speed in a horizontal direction away from the thrower and in the vertical plane containing the flag-pole and thrower. Find a relationship between the speed of projection of the stone and the speed of the bird in its flight, under the assumption that the bird is unlucky enough to be hit by the stone on its downward path.

6. The cartesian coordinates (x, y, z) of a moving particle of mass m are given at any time t, relative to an inertial frame, by the equations

$$x = ae^{-kt}, y = bt, z = ct,$$

where a, b, c and k are constants. Write down expressions for the coordinates of the particle and for the components of the velocity of the particle, each evaluated at time $t = 0$. Find an expression for the components of the angular momentum of the particle about O at time t and show that this angular momentum is in a fixed direction. Write down an expression for the components of the moment of the force acting on the particle about O.

5 • The Newtonian Model of Gravitation

Newton's universal law of gravitation is introduced, giving the gravitational force of attraction between two particles. It is shown that even bodies of finite size can be modelled as particles for the purpose of calculating their gravitational attraction, provided that they are spherically symmetric. The vertical motion of a projectile is studied, in the absence of air resistance and other non gravitational forces. The possible escape of the projectile from the earth's gravitational attraction is predicted.

5.1 Newton's Universal Law of Gravitation

Gravity, that is the attraction which draws a falling stone towards the centre of the earth, was a familiar concept long before the birth of Newton. However it was Newton who first realized that it is this same gravitational force of attraction which keeps the moon in its orbit about the earth and the planets in their orbits about the sun. As he himself wrote, 'The force which retains the celestial bodies in their orbits has been hitherto called a centripetal force; but it now being made plain that it can be no other than a gravitational force, we shall hereafter call it gravity'. Newton had deduced from the observed motions of the planets that the force which retains each planet in its orbit is directly proportional to the product of the masses of the planet and sun and inversely proportional to the square of the distance of the planet from the sun. Using the same force, but with the sun replaced by the earth, he was able to predict the observed motion of the moon. His inspiration, which he illustrated in old age by the story of the falling apple, was to realize that this inverse square force which had evaporated the remaining mysteries of celestial mechanics also applies to terrestrial mechanics, predicting the constant gravitational acceleration of a falling body.

Newton's discovery of the inverse square force dates from 1665. However one problem had to be resolved before its publication, a problem which was to cause Newton great difficulty and to delay publication for twenty years. You may have noticed that the inverse square law involves the distance between bodies of finite size. What is meant by this distance? Clearly the value of the distance will depend upon the choice of the points of the bodies between which it is measured.

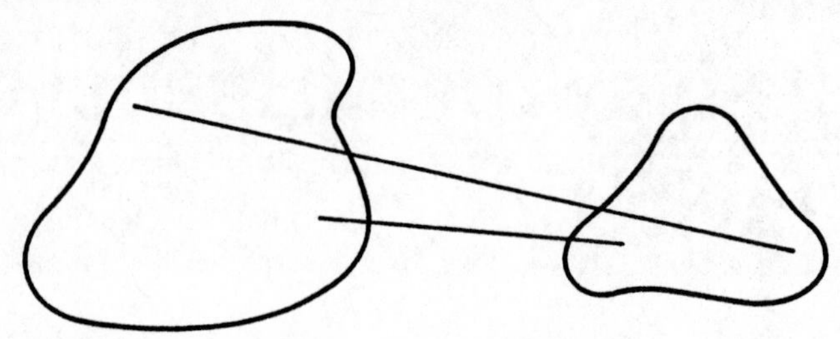

Fig 5.1 What is the distance?

TUTORIAL PROBLEM 5.1

The equatorial radii of the earth and sun are 6.378×10^6m and 6.960×10^8m, respectively, and the average distance of the earth from the sun is 1.496×10^{11}m. Calculate the maximum percentage variation in the distances between different points of the two bodies and in the inverse squares of these distances.

For some celestial bodies the variation in these measured distances will be negligible compared to the distances themselves. Then the bodies can be modelled as particles and in the next paragraph Newton's universal law of gravitation will be stated in terms of the gravitational attraction between particles. Unfortunately the example given in Tutorial Problem 5.1 shows that it is not always possible to use the relative sizes of two bodies to justify the calculation of their mutual gravitational attractions on the basis of a particle model. This then was Newton's difficulty. He resolved it by proving that the gravitational force of attraction exerted by a spherically symmetric body is the same as the gravitational force of attraction exerted by a particle of identical mass placed at the centre of the body. This result will be proved in Section 5.2 and used to model each spherically symmetric body as a particle, for the purpose of calculating the gravitational force which it exerts. In fact the planets are not quite spherically symmetric, they are slightly flattened at their poles. This flattening or oblateness leads to correction terms which have to be added to the inverse square law if the planets are to be modelled as particles. The effect of these correction terms, although significant in some contexts, is very small and throughout this chapter all bodies considered will be assumed to be spherically symmetric.

Newton's universal law of gravitation can be stated in the form

- every particle of the universe attracts every other particle with a force which is directly proportional to the product of the masses of the particles and inversely proportional to the square of the distance between them.

If the particles P and P' are located at the points with position vectors $\mathbf{r}$ and $\mathbf{r}'$ relative to a given origin O and have masses m and m' then the gravitational force $\mathbf{F}$ which the particle P' exerts on P can be written as

$$\mathbf{F} = -Gmm' \frac{(\mathbf{r} - \mathbf{r}')}{|\mathbf{r} - \mathbf{r}'|^3}.$$

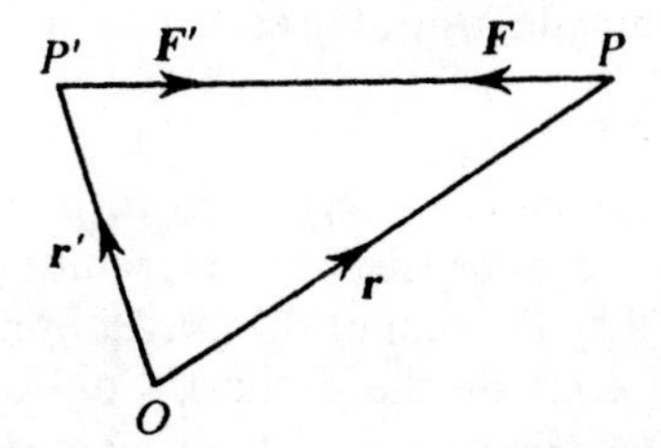

Fig 5.2 Gravitational attraction.

Here G is a positive constant of proportionality, independent of the particles, called the **universal gravitational constant.** Since $\mathbf{r} - \mathbf{r}'$ is directed from P' to P the minus sign has to be included in order that the force $\mathbf{F}$ be an attraction. As written, you may be misled into thinking that the force varies inversely as the cube of the distance $|\mathbf{r} - \mathbf{r}'|$ between the particles. However

$$|\mathbf{F}| = Gmm'\frac{|\mathbf{r} - \mathbf{r}'|}{|\mathbf{r} - \mathbf{r}'|^3} = \frac{Gmm'}{|\mathbf{r} - \mathbf{r}'|^2}$$

and the force therefore varies inversely as the square of the distance, as required.

The gravitational force $\mathbf{F}'$ which the particle P exerts on P' is obtained from $\mathbf{F}$ by interchanging $\mathbf{r}$ and $\mathbf{r}'$ so that

$$\mathbf{F}' = -Gmm'\frac{(\mathbf{r}' - \mathbf{r})}{|\mathbf{r}' - \mathbf{r}|^3}.$$

By inspection $\mathbf{F}' = -\mathbf{F}$ and so the gravitational force of attraction between two particles satisfies Newton's third law.

EXERCISES ON 5.1

1. Does the mutual gravitational attraction between two particles satisfy the strong form of Newton's third law?

2. Deduce from Example 8 of Chapter 4 the expression

$$\frac{Gmm'}{|\mathbf{r} - \mathbf{r}'|}$$

for the potential of the gravitational force which the particle P' exerts on P.

5.2 The Force of Attraction of a Spherically Symmetric Body

A body in the shape of a sphere and composed of material whose density varies only with distance from the centre of the sphere is defined to be spherically symmetric. As an example you may know that the density of the material within the earth's crust increases with depth, that is decreases with the distance from the centre of the earth, and so, neglecting the oblateness of the poles, surface irregularities, and large mineral deposits, the earth can be modelled as a spherically symmetric body.

A spherically symmetric body can be considered to be composed of concentric shells, each shell being of uniform density. The gravitational attraction which the body exerts on a particle will be the sum of the gravitational attractions which the individual concentric shells exert on the particle. Consider a uniform shell with centre O, radius a and surface density ρ units of mass per unit area. Suppose a particle of mass m is placed at the point with $OP > a$, i.e. outside the shell. Divide the shell into thin circular ribbons of width $a\delta\theta$ and radius $a \sin\theta$ as illustrated in Fig 5.3. Here $\delta\theta$ is a small increment in the angle θ. Each part of this ribbon is

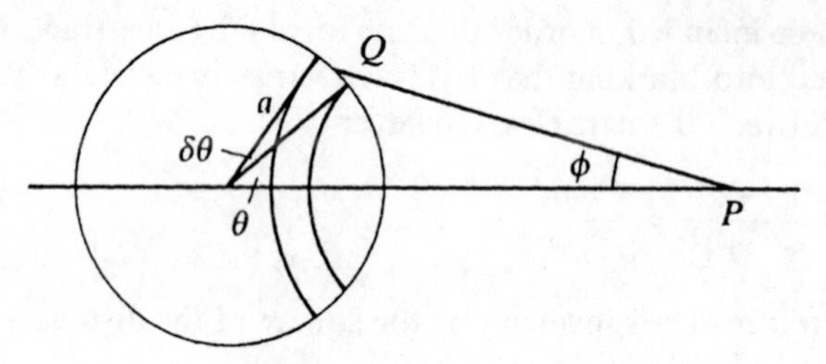

Fig 5.3 Spherical shell.

approximately the same distance PQ from P. It follows that any small mass δm_R of the ribbon will attract the particle P with a force of magnitude

$$\frac{G\delta m_R m}{PQ^2}$$

acting at an angle ϕ to the line PO. By symmetry the force with which the whole ribbon attracts the particle will be directed along PO. The component of the above force in this direction is

$$\frac{G\delta m_R m \cos\phi}{PQ^2}.$$

Summing over all the ribbon yields

$$\frac{Gm_R m \cos\phi}{PQ^2}$$

which is the force with which the ribbon attracts the particle at P. Now the area of the ribbon is $2\pi a \sin\theta\, a\delta\theta$ and therefore

$$m_R = \rho 2\pi a \sin\theta\, a\delta\theta.$$

Substituting this into the last expression yields

$$G\rho 2\pi a^2 m \frac{\sin\theta \cos\phi\, \delta\theta}{PQ^2}.$$

Integrating this from $\theta = 0$ to π will give the total gravitational force F with which the thin shell attracts the particle at P,

$$F = G\rho 2\pi a^2 m \int_{\theta=0}^{\theta=\pi} \frac{\sin\theta \cos\phi \, d\theta}{PQ^2}.$$

The three variables θ, ϕ and PQ are related by the equations

$$\cos\theta = \frac{a^2 + OP^2 - PQ^2}{2a.OP} \quad \text{and} \quad \cos\phi = \frac{OP^2 + PQ^2 - a^2}{2OP.PQ}.$$

Differentiating the first of these with respect to PQ yields

$$-\sin\theta \frac{d\theta}{dPQ} = -\frac{PQ}{aOP}.$$

In terms of the variable PQ the integral expression for F becomes

$$\begin{aligned} F &= G\rho 2\pi a^2 m \int_{PQ=OP-a}^{PQ=OP+a} \frac{PQ(OP^2 + PQ^2 - a^2)}{a.OP2OP.PQ.PQ^2} dPQ \\ &= \frac{G\rho\pi am}{OP^2} \int_{PQ=OP-a}^{PQ=OP+a} \left(1 + \frac{OP^2 - a^2}{PQ^2}\right) dPQ \\ &= \frac{G\rho\pi am}{OP^2} \left[PQ - \frac{OP^2 - a^2}{PQ}\right]_{PQ=OP-a}^{PQ=OP+a} \\ &= \frac{G\rho\pi am4a}{OP^2}. \end{aligned}$$

Now $4\pi a^2 \rho$ is the total mass M of the shell. Hence the shell attracts the particle at P with a force

$$F = \frac{GMm}{OP^2}.$$

This force of attraction is the same as that which a particle of mass M placed at O would exert on the particle at P. By adding over all the uniform concentric shells which are assumed to constitute a given spherically symmetric body the following result is obtained

- the gravitational force with which a spherically symmetric body attracts a particle placed outside the body is the same as the gravitational force with which a single particle situated at the centre of the body and of mass equal to the mass of the body would attract the particle.

This result is crucial to the modelling of celestial motions. It allows us to model any spherically symmetric body such as the sun as a particle for the purpose of calculating the gravitational force with which it attracts a particle placed outside the body.

EXERCISE ON 5.2

1. By modifying the limits on the integral in Section 5.2, prove that the gravitational force of attraction on a particle placed inside a uniform shell is zero.

5.3 Projectile Motion and the Escape Speed

Consider a particle P of mass m moving vertically. Let y be the height of the particle above the earth's surface. If the line of motion is orientated upwards then the equation of motion of the projectile is given by

$$m\ddot{y} = -\frac{GMm}{(R+y)^2},$$

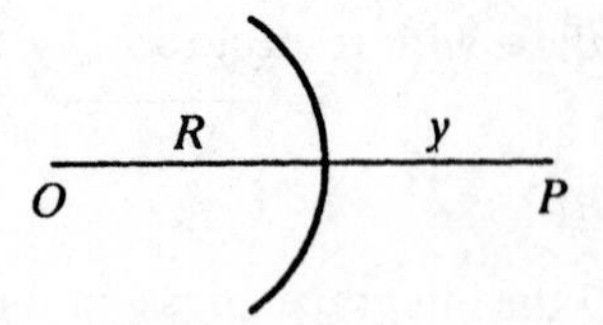

Fig 5.4 Vertical motion of a particle.

where M is the mass and R the radius of the earth; here the earth is assumed to be spherically symmetric. The mass m of the particle will cancel, but this is because we have assumed that the concept of mass appearing in Newton's second law of motion is identical to the concept of mass appearing in his universal law of gravitation. How can we justify this assumption? These two concepts of mass are called **inertial mass** and **gravitational mass**, respectively, and using subscripts i and g to distinguish between them the above equation of motion should be written as

$$m_i\ddot{y} = -\frac{GMm_g}{(R+y)^2}.$$

For a particle moving close to the earth's surface, $y \ll R$ and so the equation of motion approximates to

$$m_i\ddot{y} = -\frac{GMm_g}{R^2}$$

from which

$$\ddot{y} = -\frac{GM}{R^2}\left(\frac{m_g}{m_i}\right).$$

Here we have again obtained the constant gravitational acceleration as an approximation, this time to the inverse square force. We saw in Section 2.6 that, according to Galileo's experiment, this constant gravitational acceleration must have the same value for all particles and therefore the ratio m_g/m_i must be a constant, independent of the particle P. This result is known as the equivalence of gravitational and inertial mass and it explains why Galileo's experiment plays such a central role in mathematical models of gravitation. We have not specified the unit of gravitational mass and we are free to choose it to make the ratio m_g/m_i equal to one; this choice will dictate the actual value of the universal constant G. Then $m_g = m_i$ and there is no need to distinguish between the two concepts of mass.

Furthermore the gravitational acceleration g is then related to the mass and radius of the earth through the equation

$$g = \frac{GM}{R^2}.$$

Example 12 of Chapter 2 shows that under the assumption of a constant gravitational acceleration any particle projected upwards will attain a maximum height and then fall back to earth. Let us now investigate the motion of such a projectile more accurately by introducing the inverse square force and the equation of motion written down at the beginning of this section. Cancelling the mass m of the projectile gives

$$\ddot{y} = -\frac{GM}{(R+y)^2}.$$

Multiplying this equation by the velocity $\dot{y}$ and integrating with respect to time yields

$$\tfrac{1}{2}\dot{y}^2 = \frac{GM}{R+y} + \text{constant}.$$

This equation expresses the conservation of energy per unit mass of the projectile and can be written in the usual form

$$\tfrac{1}{2}\dot{y}^2 + V(y) = E,$$

where

$$V(y) = -\frac{GM}{R+y}$$

is the potential energy per unit mass of the projectile. Incidentally this potential energy and therefore the energy equation itself could have been written down directly using the result obtained in question 2 of Exercises 5.1. The graph of $V(y)$ against y is sketched in Fig 5.5. From this it can be seen, using the methods first introduced in Section 2.5, that if $E < 0$ the height of the projectile will increase until the line $V = E$ meets the graph. The direction of motion then reverses and the projectile falls back to earth. This motion is qualitatively the same as motion under a constant gravitational acceleration. However if $E \geq 0$ the projectile will escape to

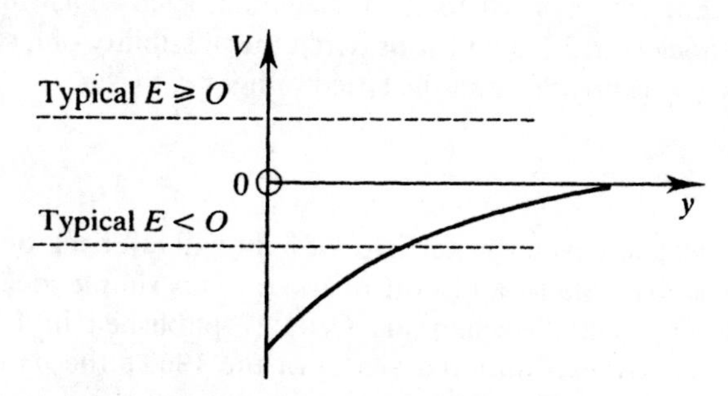

Fig 5.5 Potential energy of projectile.

infinity. In this case we say that the projectile escapes the gravitational attraction of the earth.

Consider a projectile which is launched from the earth's surface $y = 0$ with velocity $\dot{y} = v_0 > 0$. Substituting these initial conditions into the energy equation gives

$$\tfrac{1}{2}v_0^2 - \frac{GM}{R} = E.$$

The condition $E \geq 0$ for the projectile to escape can now be written as

$$v_0^2 \geq \frac{2GM}{R}$$

or, remembering that $v_0 > 0$,

$$v_0 \geq \sqrt{\frac{2GM}{R}}.$$

The critical speed $\sqrt{2GM/R}$ is called the **escape speed** and is denoted by v_e. Hence

$$v_e = \sqrt{\frac{2GM}{R}}$$

or, in terms of g,

$$v_e = \sqrt{2gR}.$$

This analysis of projectile motion would be equally valid for any other planet or star provided that M and R are reinterpreted as the mass and radius of that planet or star. For the earth $M = 5.976 \times 10^{24}$kg and $R = 6.378 \times 10^6$m. Taking $G = 6.670 \times 10^{-11}\text{m}^3\text{kg}^{-1}\text{s}^{-2}$ leads to the value $11.2 \times 10^3\text{ms}^{-1}$ for the escape speed. This is about 30 times the speed of sound travelling in air at standard temperature and pressure.

According to Newton's universal law of gravitation only particles possessing mass will experience a gravitational force so that, for example, the gravitational force experienced by a massless particle of light, that is a photon, will be zero. Nevertheless, we saw that the mass cancels in the equation of motion leaving

$$\ddot{y} = -\frac{GM}{(R+y)^2}$$

and it is tempting to assume that this equation holds for both massive and massless particles. An English physicist John Michell made such an assumption in a lecture to the Royal Society in 1783 and considered the possibility of a star existing whose escape velocity v_e is greater than the speed of light c, that is

$$\sqrt{\frac{2GM}{R}} > c.$$

With this assumption even the particles of light will fall back onto the star and so the star will be invisible to a distant observer. This simple idea also appeared in Pierre Laplace's book "Mechanique Celeste" published in 1793 but was then forgotten for nearly two hundred years. In the 1960's the name **black hole** was coined for such stars. No one accepts the validity of the above argument –

Newton's law really does imply that the gravitational force exerted on a massless particle is zero. However in Einstein's model of gravitation, his theory of general relativity, the motion of a massless particle is influenced by gravitation. This theory predicts the existence of black holes and it is a curious fact that the theory leads to the same inequality as obtained using the false argument, by Michell and Laplace! From this inequality the radius of a black hole must satisfy

$$R < \frac{2GM}{c^2}.$$

Consider a black hole of mass equal to the mass of the sun. Then $M = 1.990 \times 10^{30}$kg and taking $G = 6.670 \times 10^{-11} m^3 \text{kg}^{-1} \text{s}^{-2}$ and $c = 2.998 \times 10^8 \text{ms}^{-1}$ leads to the conclusion that the radius of the black hole must be less than 2.95×10^3m!

Example I

A particle is projected vertically from a point on the surface of the earth. Show that if it does not escape the earth's gravitational attraction then it will attain its greatest height after a time T given by

$$T = \sqrt{\frac{R+H}{2GM}}\left[\sqrt{RH} + (R+H)\cos^{-1}\sqrt{\frac{R}{R+H}}\right].$$

SOLUTION

Consider the energy equation

$$\tfrac{1}{2}\dot{y}^2 - \frac{GM}{R+y} = E.$$

The greatest height $y = H$ is achieved when the velocity $\dot{y}$ becomes zero and therefore

$$-\frac{GM}{R+H} = E.$$

Eliminating E and rearranging gives

$$\dot{y}^2 = \frac{2GM}{R+H}\frac{H-y}{R+y}.$$

Before the particle reaches its greatest height its velocity $\dot{y}$ will always be positive. It follows that

$$\frac{dy}{dt} = \sqrt{\frac{2GM}{R+H}}\frac{\sqrt{H-y}}{\sqrt{R+y}}$$

so that

$$\int_{y=0}^{H}\frac{\sqrt{R+y}}{\sqrt{H-y}}dy = \sqrt{\frac{2GM}{R+H}}\int_{t=0}^{T}dt.$$

To evaluate the integral I on the left hand side of this equation, put $y = x^2 - R$, with $x \geq 0$, so that

$$I = \int_{x=\sqrt{R}}^{\sqrt{R+H}} \frac{2x^2}{\sqrt{R+H-x^2}}\,dx.$$

Throughout the range of integration $x^2 \le R + H$ and so it is legitimate to make the further substitution

$$x = \sqrt{R+H}\cos\theta.$$

Then

$$I = -(R+H)\int_{\theta=\theta_0}^{0} 2\cos^2\theta\, d\theta,$$

where $\theta_0 = \cos^{-1}\sqrt{R/(R+H)}$. Hence

$$I = -(R+H)\int_{\theta=\theta_0}^{0} (\cos 2\theta + 1)d\theta = (R+H)(\tfrac{1}{2}\sin 2\theta_0 + \theta_0).$$

Now

$$\tfrac{1}{2}\sin 2\theta_0 = \sin\theta_0 \cos\theta_0 = \sqrt{\frac{R}{R+H}}\sqrt{1-\frac{R}{R+H}} = \frac{\sqrt{RH}}{R+H}$$

and so

$$I = \sqrt{RH} + (R+H)\cos^{-1}\sqrt{\frac{R}{R+H}}.$$

Substituting this into the original equation, and noting that $\int_{t=0}^{T} dt = T$, gives the required answer, namely

$$T = \sqrt{\frac{R+H}{2GM}}\left[\sqrt{RH} + (R+H)\cos^{-1}\sqrt{\frac{R}{R+H}}\right].$$

Notice that if H is small compared to R then $\cos^{-1}\sqrt{R/(R+H)} \approx \cos^{-1} 1 = 0$. Hence

$$T \approx \sqrt{\frac{R^2 H}{2GM}}.$$

In terms of $g = GM/R^2$ this can be written as

$$T = \sqrt{H/2g},$$

a result which could have been obtained directly by modelling the gravitational attraction of the earth as a constant gravitational acceleration. •

EXERCISES ON 5.3

1. The mass and radius of the moon are 7.36×10^{22}kg and 1.74×10^{6}m, respectively. Calculate the gravitational acceleration and the escape speed at the surface of the moon.

2. At what height must a projectile be launched in order that its escape speed be one half of the escape speed at the surface of the earth?

3. A particle is projected vertically from a point on the earth's surface with a speed $v_0 < v_e$. Prove that the maximum height attained is

$$R^2v_0^2/(2GM - Rv_0^2),$$

where R and M are the radius and mass of the earth respectively.

4. Show that if the maximum height in the last question is small compared to the radius of the earth then it is approximately equal to $v_0^2/2g$, where g is the gravitational acceleration. Devise a question, suitable for Section 2.6, which would give the answer $v_0^2/2g$ exactly.

5. A particle is released from rest from a height H. Show that it will hit the earth with a speed equal to

$$\sqrt{\frac{2GMH}{R(R+H)}},$$

where R is the radius and M the mass of the earth. Find an expression for the time of fall.

Summary

- the equation $\mathbf{F} = -Gmm'(\mathbf{r} - \mathbf{r}')/|\mathbf{r} - \mathbf{r}'|^3$ giving the force with which a particle of mass m' at $\mathbf{r}'$ attracts a particle of mass m at $\mathbf{r}$ is an expression of **Newton's universal law of gravitation**; the force is directly proportional to the product of the masses and inversely proportional to the square of the distance between the particles
- G is the **universal gravitational constant**
- the gravitational attraction of a spherically symmetric body is identical to that of a particle situated at the centre of the body and of mass equal to the mass of the body
- Galileo's experiment from the Leaning Tower of Pisa established the equivalence of **inertial** and **gravitational** mass
- the constant **gravitational acceleration** g and **escape speed** v_e for a spherically symmetric body of mass M and radius R are given by $g = GM/R^2$ and $v_e = \sqrt{2GM/R}$

FURTHER EXERCISES

1. Using the data given in Tutorial Problem 5.1 calculate the maximum percentage variation in the distance between different points of the earth and the centre of the sun. Would your answer justify modelling the earth as a particle regardless of whether or not it is spherically symmetric?

2. Prove that the gravitational force exerted by a uniform solid sphere on a particle within the sphere is proportional to the distance of the particle from the centre of the sphere. A hole is drilled through such a sphere, passing through its centre.

A particle is released from rest at one end of the hole. Discuss the motion of the particle as it moves under the attraction of the sphere alone.

3. A particle is projected vertically from a point on the earth's surface with a velocity which is such that, if the gravitational effect of the earth were modelled as a uniform acceleration, the maximum height of the particle would be H_0. Show that the maximum height actually attained is greater than H_0 by an amount $H_0^2/(R - H_0)$ where R is the radius of the earth.

4. A particle is released from rest from a height H. How far will it have fallen when it attains one half of its final speed?

   (a) under a uniform gravitational attraction and

   (b) under an inverse square gravitational force?

   Obtain the answer for (a) as an approximation to the answer for (b).

5. Show that the gravitational acceleration at the surface of a black hole having the same mass as the sun is 10^{11} times greater than the gravitational acceleration at the surface of the sun itself. Discuss whether a constant gravitational acceleration could provide a viable mathematical model for the gravitational attraction of a black hole.

6 • Circular Motion

The radial and transverse components of velocity and of acceleration are obtained using plane polar coordinates and the results are applied to the motion of a particle moving on a circle fixed in an inertial frame. The case of motion on a circle with constant angular velocity is discussed in detail from the point of view of a non inertial frame of reference in which the particle is at rest. This leads to the general theory of rotating frames which is then applied to the motion of a particle relative to the rotating earth.

6.1 Plane Polar Coordinates

As you saw in Section 4.1, the use of the cartesian basis associated with the axes specifying a given frame of reference S with origin O is particularly convenient when differentiating vectors relative to that frame. The derivative of a vector is found by differentiating the cartesian components of the vector. In particular the velocity of a particle is found by differentiating the cartesian coordinates of the particle, this result follows from the fact that the cartesian components of the position vector of a point are the cartesian coordinates of that point. Despite the ease with which derivatives are found using cartesians it is sometimes advantageous, for specific problems, to use other coordinates and care has then to be taken in calculating derivatives and in particular in finding expressions for the velocity and acceleration of a particle. In this Chapter we will illustrate the use of other coordinates and their associated bases by discussing the motion of a particle moving on a fixed plane in terms of plane polar coordinates.

Consider the motion of a particle P, moving on a plane, relative to a given origin O and frame of reference S specified by the cartesian axes illustrated in Fig. 6.1. The

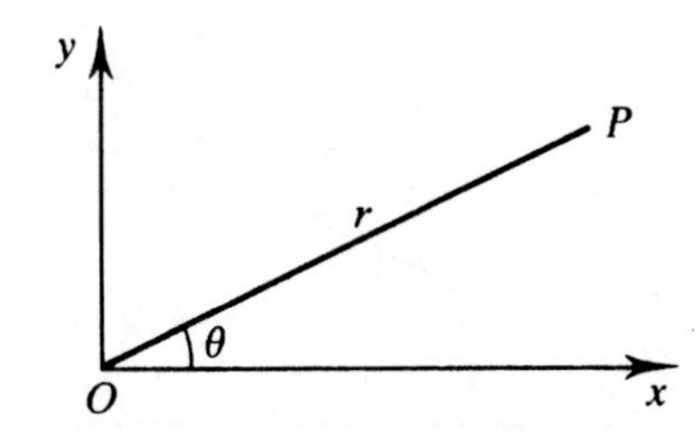

Fig 6.1 Plane cartesian and polar coordinates.

figure also illustrates the familiar plane polar coordinates (r, θ) of P relative to the same origin and with the initial line $\theta = 0$ chosen to coincide with the positive x axis. These polar coordinates are related to the cartesian coordinates (x, y) of P by the equations

$$x = r \cos \theta, \quad y = r \sin \theta.$$

Squaring these and adding gives $x^2 + y^2 = r^2$ and r, being positive, is found by taking the positive square root. Assuming that $r \neq 0$, expressions for $\cos\theta$ and $\sin\theta$ are obtained by substituting r into the two original equations. These expressions will only determine a unique value for θ if the angle is constrained to lie in an appropriate interval. Here we shall use the convention $0 \leq \theta < 2\pi$. Each point of the plane is then specified uniquely by the values of the coordinates (r, θ), with the exception of the origin O for which the angle θ is undefined.

The basis vectors $\hat{\mathbf{i}}$ and $\hat{\mathbf{j}}$ associated with the cartesian coordinates (x, y) are often defined as being the unit vectors parallel to Ox and Oy and directed in the positive x and y directions, respectively. This definition does not generalize to other coordinates and we prefer to define the basis vectors at each point P, $\hat{\mathbf{i}}$ being the unit vector in the direction in which P moves when x is increased keeping y fixed and $\hat{\mathbf{j}}$ being the unit vector in the direction in which P moves when y is increased keeping x fixed. These basis vectors are illustrated in Fig. 6.2. This definition

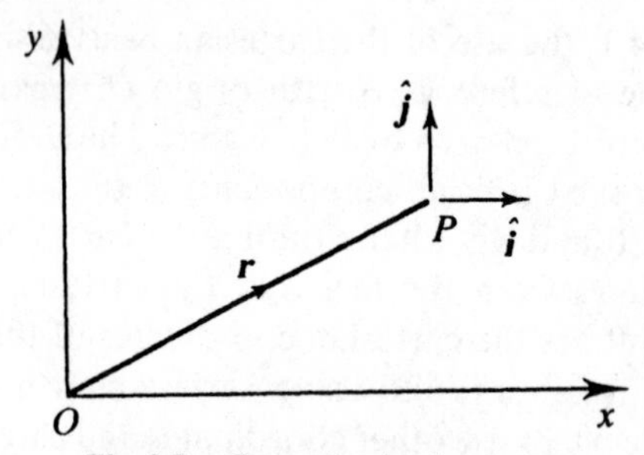

Fig 6.2 Cartesian basis vectors.

generalizes to other coordinates. In particular, basis vectors $\hat{\mathbf{r}}$ and $\hat{\boldsymbol{\theta}}$ can be associated with the plane polar coordinates (r, θ) at each point P, $\hat{\mathbf{r}}$ being the unit vector in the direction in which P moves when r is increased keeping θ fixed and $\hat{\boldsymbol{\theta}}$ being the unit vector in the direction in which P moves when θ is increased keeping r fixed. These basis vectors are illustrated in Fig 6.3. The vectors $\hat{\mathbf{r}}$ and $\hat{\boldsymbol{\theta}}$ constitute a

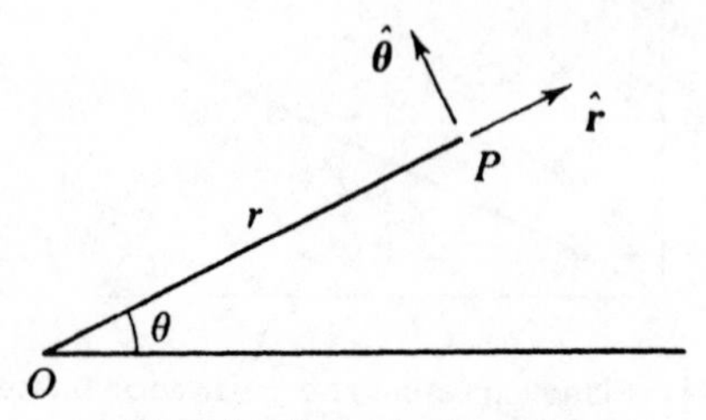

Fig 6.3 Polar basis vectors.

polar basis and the components of a given vector relative to this basis are called the **polar components**. The directions of $\hat{\mathbf{r}}$ and $\hat{\boldsymbol{\theta}}$ are often described by the adjectives **radial** and **transverse**, respectively, and the polar components of the vector are then referred to as the **radial** and **transverse components**. You can see from the above two figures that

$$\mathbf{r} = \overrightarrow{OP} = x\hat{\mathbf{i}} + y\hat{\mathbf{j}} \quad \text{and} \quad \mathbf{r} = \overrightarrow{OP} = r\hat{\mathbf{r}}.$$

Notice that the polar components of $\mathbf{r}$ are $(r, 0)$ and are not equal to the polar coordinates of P.

The fact that both of the above bases, cartesian and polar, are orthonormal makes it very easy to write down the components of a given vector relative to either. Each component is found by multiplying the magnitude of the vector by the cosine of the angle between the vector and the appropriate basis vector. In particular the polar basis vectors can be written in terms of the cartesian basis vectors as

$$\hat{\mathbf{r}} = \cos\theta\,\hat{\mathbf{i}} + \sin\theta\,\hat{\mathbf{j}}$$

$$\text{and} \quad \hat{\boldsymbol{\theta}} = -\sin\theta\,\hat{\mathbf{i}} + \cos\theta\,\hat{\mathbf{j}}.$$

This is best seen by superimposing Fig 6.2 onto Fig 6.3; incidently, this also demonstrates that the relationship can be described as an anticlockwise rotation through an angle θ.

TUTORIAL PROBLEM 6.1

Invert the equations relating $\hat{\mathbf{r}}$ and $\hat{\boldsymbol{\theta}}$ to $\hat{\mathbf{i}}$ and $\hat{\mathbf{j}}$ algebraically. Notice that the equations you obtain can be written down directly from the original equations by interchanging $\hat{\mathbf{r}}$ and $\hat{\boldsymbol{\theta}}$ with $\hat{\mathbf{i}}$ and $\hat{\mathbf{j}}$, and replacing the angle θ by $-\theta$. Explain this.

Example 1

The vector $\mathbf{v}$, defined at the point P with cartesian coordinates $(-1, 2)$, has cartesian components (3,4). Find the polar components of $\mathbf{v}$.

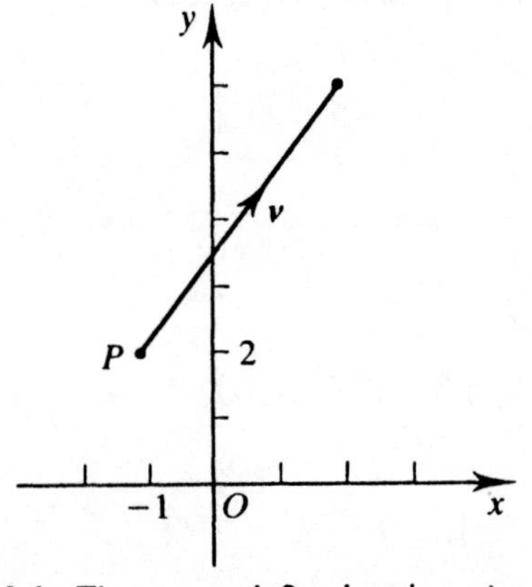

Fig 6.4 The vector defined at the point P.

SOLUTION

We are given that

$$\mathbf{v} = 3\hat{\mathbf{i}} + 4\hat{\mathbf{j}}.$$

In order to obtain the polar components of $\mathbf{v}$, the vector must be written as a linear combination of $\hat{\mathbf{r}}$ and $\hat{\boldsymbol{\theta}}$. This is done by using the equations

$$\hat{\mathbf{i}} = \cos\theta\,\hat{\mathbf{r}} - \sin\theta\,\hat{\boldsymbol{\theta}} \quad \text{and} \quad \hat{\mathbf{j}} = \sin\theta\,\hat{\mathbf{r}} + \cos\theta\,\hat{\boldsymbol{\theta}}$$

obtained in Tutorial Problem 6.1, the angle θ being the polar coordinate of the point P at which the vector is defined. Now

$$-1 = r\cos\theta \quad \text{and} \quad 2 = r\sin\theta$$

so that $r = \sqrt{5}$. Substituting back for r gives

$$\cos\theta = -1/\sqrt{5} \quad \text{and} \quad \sin\theta = 2/\sqrt{5}.$$

Hence

$$\hat{\mathbf{i}} = -\frac{1}{\sqrt{5}}\hat{\mathbf{r}} - \frac{2}{\sqrt{5}}\hat{\boldsymbol{\theta}} \quad \text{and} \quad \hat{\mathbf{j}} = \frac{2}{\sqrt{5}}\hat{\mathbf{r}} - \frac{1}{\sqrt{5}}\hat{\boldsymbol{\theta}}.$$

The vector $\mathbf{v}$ can now be rewritten as

$$\mathbf{v} = 3\left(-\frac{1}{\sqrt{5}}\hat{\mathbf{r}} - \frac{2}{\sqrt{5}}\hat{\boldsymbol{\theta}}\right) + 4\left(\frac{2}{\sqrt{5}}\hat{\mathbf{r}} - \frac{1}{\sqrt{5}}\hat{\boldsymbol{\theta}}\right) = \frac{5}{\sqrt{5}}\hat{\mathbf{r}} - \frac{10}{\sqrt{5}}\hat{\boldsymbol{\theta}}.$$

The required polar components of $\mathbf{v}$ are $(\sqrt{5}, -2\sqrt{5})$. •

The velocity $\mathbf{v}$ of the particle P, relative to the frame of reference S and origin O introduced above, is

$$\mathbf{v} = \dot{x}\hat{\mathbf{i}} + \dot{y}\hat{\mathbf{j}},$$

found by differentiating the position vector $\mathbf{r} = x\hat{\mathbf{i}} + y\hat{\mathbf{j}}$ with respect to time. The velocity can also be found by differentiating $\mathbf{r} = r\hat{\mathbf{r}}$ with respect to time. You will notice from Fig 6.3 that as the particle P moves the direction of the polar basis vectors will, in general, change so that $\hat{\mathbf{r}}$ and $\hat{\boldsymbol{\theta}}$ are themselves functions of time t. The derivatives of these functions are easily found. Using the chain rule

$$\frac{d\hat{\mathbf{r}}}{dt} = \frac{d\theta}{dt}\frac{d\hat{\mathbf{r}}}{d\theta}$$

to differentiate

$$\hat{\mathbf{r}} = \cos\theta\,\hat{\mathbf{i}} + \sin\theta\,\hat{\mathbf{j}}$$

with respect to time gives

$$\dot{\hat{\mathbf{r}}} = \dot{\theta}(-\sin\theta\,\hat{\mathbf{i}} + \cos\theta\,\hat{\mathbf{j}}) = \dot{\theta}\hat{\boldsymbol{\theta}}.$$

Similarly

$$\dot{\hat{\boldsymbol{\theta}}} = -\dot{\theta}\hat{\mathbf{r}}.$$

Now

$$\begin{aligned}\mathbf{v} = (\dot{r\hat{\mathbf{r}}}) &= \dot{r}\hat{\mathbf{r}} + r\dot{\hat{\mathbf{r}}} \\ &= \dot{r}\hat{\mathbf{r}} + r\dot{\theta}\hat{\boldsymbol{\theta}}.\end{aligned}$$

The polar components of the velocity are therefore $(\dot{r}, r\dot{\theta})$, $\dot{r}$ being the radial component and $r\dot{\theta}$ the transverse component. Similarly the acceleration $\mathbf{a}$ is given by

$$\begin{aligned}\mathbf{a} = \dot{\mathbf{v}} &= (\dot{r}\hat{\mathbf{r}} + r\dot{\theta}\hat{\boldsymbol{\theta}}) \\ &= \ddot{r}\hat{\mathbf{r}} + \dot{r}\dot{\hat{\mathbf{r}}} + \dot{r}\dot{\theta}\hat{\boldsymbol{\theta}} + r\ddot{\theta}\hat{\boldsymbol{\theta}} + r\dot{\theta}\dot{\hat{\boldsymbol{\theta}}} \\ &= \ddot{r}\hat{\mathbf{r}} + 2\dot{r}\dot{\theta}\hat{\boldsymbol{\theta}} + r\ddot{\theta}\hat{\boldsymbol{\theta}} - \dot{r}\dot{\theta}^2\hat{\mathbf{r}} \\ &= (\ddot{r} - r\dot{\theta}^2)\hat{\mathbf{r}} + (r\ddot{\theta} + 2\dot{r}\dot{\theta})\hat{\boldsymbol{\theta}}.\end{aligned}$$

The polar components of the acceleration are therefore $(\ddot{r} - r\dot{\theta}^2, r\ddot{\theta} + 2\dot{r}\dot{\theta})$, $\ddot{r} - \dot{r}\dot{\theta}^2$ being the radial component and $r\ddot{\theta} + 2\dot{r}\dot{\theta}$ being the transverse component.

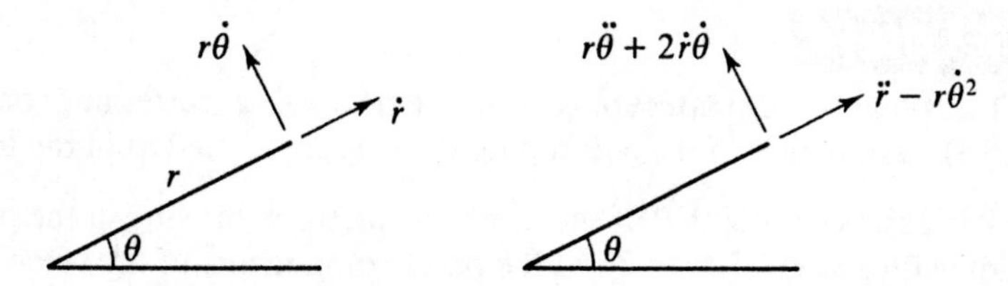

Fig 6.5 Radial and transverse components of velocity and acceleration.

TUTORIAL PROBLEM 6.2

Check that the expressions for the radial and transverse components of velocity and acceleration can be obtained by differentiating the equation $x = r\cos\theta$ twice with respect to time and evaluating the resulting expressions for $\dot{x}$ and $\ddot{x}$ when $\theta = 0$ and $\theta = 3\pi/2$. Can you explain why?

Example 2

The equation $r = a(2 + \cos\theta)$ is the polar equation of a curve known as the limacon of Pascal, illustrated overleaf. A particle describes this curve, moving in the sense of increasing θ with a constant speed v. Show that

$$\dot{\theta} = v/a\sqrt{5 + 4\cos\theta}.$$

SOLUTION

The speed v of the particle is the magnitude of the velocity $\dot{r}\hat{\mathbf{r}} + r\dot{\theta}\hat{\boldsymbol{\theta}}$ so that

$$v^2 = \dot{r}^2 + r^2\dot{\theta}^2.$$

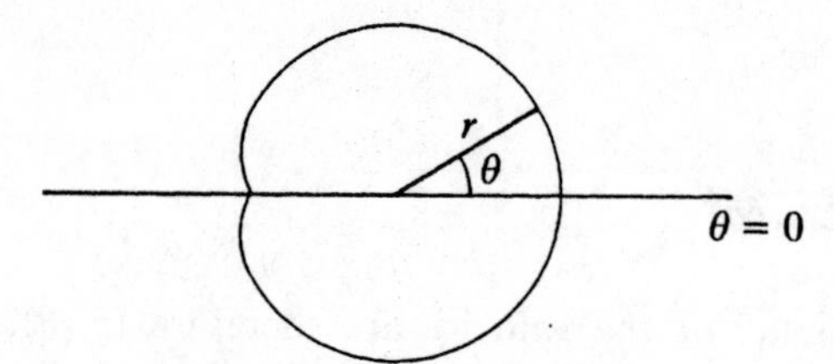

Fig 6.6 Limacon of Pascal.

Differentiating $r = a(2 + \cos\theta)$ with respect to time yields

$$\dot{r} = -a\sin\theta\,\dot{\theta}$$

so that

$$\begin{aligned} v^2 &= a^2\sin^2\theta\,\dot{\theta}^2 + a^2(2+\cos\theta)^2\dot{\theta}^2 \\ &= a^2(5 + 4\cos\theta)\dot{\theta}^2. \end{aligned}$$

Since $\dot{\theta} > 0$ it follows that $\dot{\theta} = v/a\sqrt{5 + 4\cos\theta}$. ●

EXERCISES ON 6.1

1. Find the polar coordinates of the point P having cartesian coordinates $(-2.6, 5.3)$, assuming that the coordinate systems are related as in the text.

2. A particle passes the point P of the previous question, moving in the positive x direction with speed 4.7ms^{-1}. Find the polar components of the velocity of the particle as it passes P.

3. Prove that the speed of a particle moving on the path $r = a\cos\theta$ is constant if and only if $\dot{\theta}$ is constant.

4. Show that the expression for the transverse component of the acceleration can be written as $\frac{1}{r}\frac{d}{dt}(r^2\dot{\theta})$.

6.2 Circular Motion

The position of a particle P moving on a fixed circle of radius r, centred at O, is best specified by the directed angle θ between OP and a fixed initial line l passing through O. To define such an angle it is necessary first to choose an orientation of the circle which can be indicated in diagrams by an arrowhead as in Fig 6.7(i). The directed angle θ is then taken to be positive if the rotation from l to OP is in the sense of the orientation and negative if the rotation is in the opposite sense. Perhaps you can see the analogy with motion on a straight line? Incidentally, the orientation of the circle can also be indicated as in Fig 6.7(ii). The directed angle θ is called the **angular displacement** of P relative to l. For a moving particle, the angular displacement will be a function of time and its derivative $\dot{\theta}$ is called the **angular velocity of P relative to l.** The sign of the angular velocity indicates the direction of motion of the particle; if $\dot{\theta} > 0$ the particle is moving in the direction in which the circle is orientated, and if $\dot{\theta} < 0$ the particle is moving in the negative

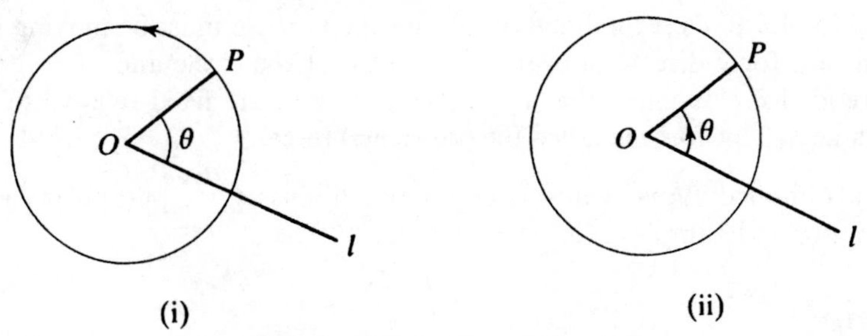

Fig 6.7 Orientation of the circle.

direction. If $\dot{\theta} = 0$ at some instant of time then the particle is stationary or instantaneously at rest at that time.

TUTORIAL PROBLEM 6.3

Would you agree with the statement that if $\dot{\theta} > 0$ the particle P in Fig 6.7 is moving anticlockwise? You should compare your conclusions with the discussion given immediately after Fig 2.2.

Plane polar coordinates with origin O and initial line l are particularly well adapted to the discussion of circular motion. The angular displacement is chosen as the polar coordinate θ, the polar coordinate r being constant and equal to the radius of the circle; note that the interval over which θ is defined will now depend on the particular problem being discussed. The expressions for the velocity and acceleration obtained in Section 6.1 then reduce to

$$\mathbf{v} = r\dot{\theta}\hat{\boldsymbol{\theta}} \quad \text{and} \quad \mathbf{a} = -r\dot{\theta}^2\hat{\mathbf{r}} + r\ddot{\theta}\hat{\boldsymbol{\theta}}.$$

The basis vectors $\hat{\mathbf{r}}$ and $\hat{\boldsymbol{\theta}}$ and the radial and transverse components of the velocity and acceleration of the particle P are illustrated in Fig 6.8. Notice that the speed $|\mathbf{v}|$ of the particle is equal to $r|\dot{\theta}|$. Consider now the special case of circular motion with constant angular velocity. Then $\ddot{\theta} = 0$ and the acceleration of the particle relative to the centre of the circle is radial. This radial acceleration is directed towards the centre of the circle and has magnitude $r\dot{\theta}^2$. In applications the angular velocity is usually denoted by ω and the magnitude of the radial acceleration is then $r\omega^2$.

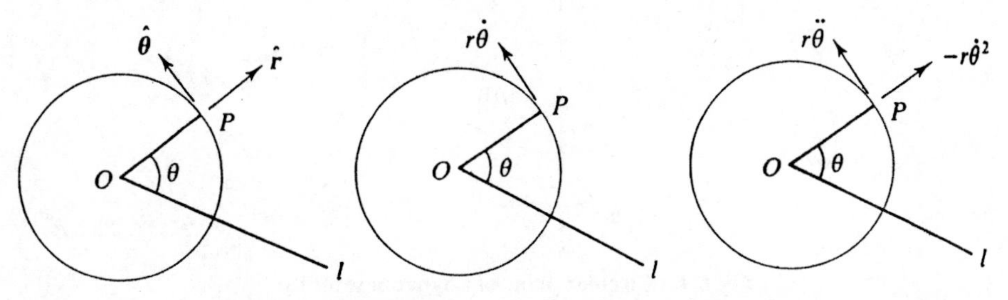

Fig 6.8 Velocity and acceleration for circular motion.

According to Newton's second law of motion the particle must be moving under the action of a force directed towards the centre of the circle and of magnitude $mr\omega^2$; here we have assumed that the circle and line l are fixed relative to some inertial frame S. This force is called the **centripetal force.**

The results obtained above will now be used to discuss some particular cases of circular motion in Examples 3–6.

Example 3

Syncom satellites.

Synchronized communication satellites are placed in orbits chosen so that the satellites appear to be stationary relative to each point on the earth's surface. The advantage of this is that signals can be bounced off such a satellite without the need to track it and continuously adjust the bearings of the transmitter and receiver. Show that the radius of the orbit of such a satellite is 4.22×10^7m.

SOLUTION

In order that a given Syncom satellite should appear stationary relative to each point on the earth's surface, its orbit must be circular and lie in a plane perpendicular to the axis of the earth. Furthermore, the rotation of the satellite in this orbit must be synchronized with the rotation of the earth about its axis. The angular velocity ω of the satellite in its circular orbit must therefore be constant and equal to one rotation per day, i.e.

$$\omega = \frac{2\pi}{24 \times 60 \times 60} = 7.27 \times 10^{-5}\text{rads}^{-1}.$$

The force constraining the satellite to move on its circular orbit is the gravitational force of attraction of the earth, i.e.

$$-\frac{GMm}{R^2}\hat{\mathbf{R}},$$

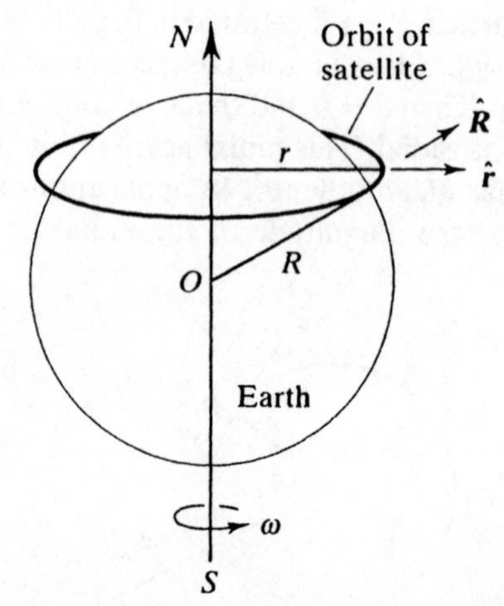

Fig 6.9 Circular orbit of a Syncom satellite.

where m is the mass of the satellite, M is the mass of the earth, R is the distance between the satellite and the centre of the earth and $\hat{\mathbf{R}}$ is a unit vector directed radially away from the centre of the earth, see Fig 6.9. The centripetal force is

$$-mr\omega^2\hat{\mathbf{r}},$$

where r is the radius of the circular orbit and $\hat{\mathbf{r}}$ is a unit vector directed radially away from the centre of the orbit. The centripetal force and gravitational force are one and the same so that

$$-mr\omega^2\hat{\mathbf{r}} = -\frac{GMm}{R^2}\hat{\mathbf{R}},$$

an equation which could equally well be written down by considering the acceleration of the satellite and using Newton's second law of motion. This equation requires that $\hat{\mathbf{R}} = \hat{\mathbf{r}}$ so that the orbit of the satellite must lie in the equatorial plane. Then $R = r$ and

$$mr\omega^2 = \frac{GMm}{r^2}$$

which gives $r = \sqrt[3]{GM/\omega^2}$. Taking $M = 5.976 \times 10^{24}\,\text{kg}$, $G = 6.670 \times 10^{-11}\,\text{m}^3\text{kg}^{-1}\text{s}^{-1}$ and $\omega = 7.27 \times 10^{-5}\,\text{rad s}^{-1}$ leads to

$$r = 4.22 \times 10^7\,\text{m}.$$

•

In this Example the gravitational attraction has been written down on the supposition that both the earth and the satellite are modelled as particles. Assuming spherical symmetry the earth can indeed by modelled as a particle. However a communication satellite is far from spherically symmetric and it is necessary to make a case for modelling it as a particle based on its size; the size of a communication satellite is indeed small compared to the radius of its orbit and so can be neglected.

Example 4

The simple pendulum.

A simple pendulum consists of a particle swinging in a fixed vertical plane at one end of a light inextensible string, the other end of which is attached to a fixed point. Show that if the amplitude of the swing is small then the motion of the particle is simple harmonic and obtain an expression for the frequency.

SOLUTION

The pendulum is illustrated in Fig 6.10, the angular displacement θ of the particle being taken relative to the downward vertical. The mass of the particle and length of the string are denoted by m and l respectively. The acceleration of the particle P is given by

$$\mathbf{a} = -l\dot{\theta}^2\hat{\mathbf{r}} + l\ddot{\theta}\hat{\boldsymbol{\theta}}.$$

Neglecting air resistance, the only forces acting on P are the tension T and weight mg as illustrated. The resultant force is therefore

$$\mathbf{F} = -T\hat{\mathbf{r}} + mg\cos\theta\,\hat{\mathbf{r}} - mg\sin\theta\,\hat{\boldsymbol{\theta}}.$$

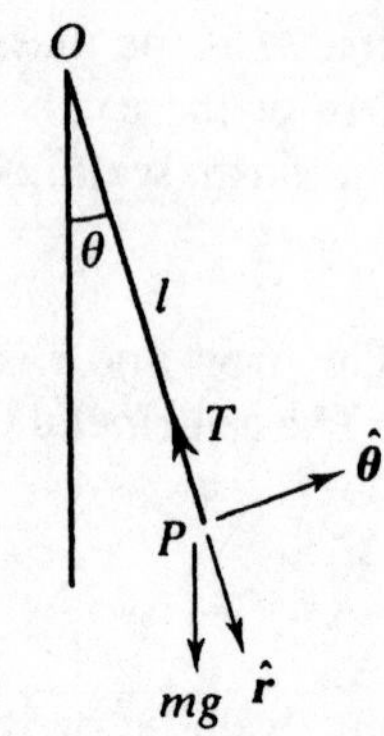

Fig 6.10 The simple pendulum.

The transverse component of the equation of motion $m\mathbf{a} = \mathbf{F}$ is

$$ml\ddot{\theta} = -mg \sin \theta$$

so that

$$\ddot{\theta} + \frac{g}{l} \sin \theta = 0.$$

If the particle is swinging with a small amplitude then the angle θ is small and $\sin \theta$ approximates to θ giving

$$\ddot{\theta} + \frac{g}{l} \theta = 0.$$

This is the simple harmonic motion equation with frequency

$$\frac{1}{2\pi}\sqrt{\frac{g}{l}}.$$

●

Although the simple pendulum is defined in terms of a particle and light inextensible string it often provides a good mathematical model for other, more realistic mechanical systems. Notice that an appropriate interval for θ in the above Example is $-\pi \leq \theta < \pi$; this ensures that θ is continuous as the string passes the downward vertical.

Example 5

Motion on a complete vertical circle.

Consider a particle P of mass m hanging at rest at the end of a light inextensible string of length l, the other end of which is attached to a fixed point O. The particle is given an initial angular velocity ω. Find conditions on ω for the particle to complete a vertical circle, air resistance being neglected.

SOLUTION

The situation is again illustrated in Fig 6.10 and the two components of the equation of motion are

$$ml\ddot{\theta} = -mg \sin \theta$$

$$\text{and} \quad -ml\dot{\theta}^2 = -T + mg \cos \theta.$$

Multiplying the transverse component of the equation of motion by the angular velocity $\dot{\theta}$ and integrating with respect to time gives

$$ml \int \ddot{\theta}\dot{\theta} dt = -mg \int \sin \theta \, \dot{\theta} dt$$

$$\text{so that} \quad ml \int \dot{\theta} d\dot{\theta} = -mg \int \sin \theta \, d\theta.$$

Integrating and rearranging gives the energy equation in the form

$$\frac{1}{2}ml\dot{\theta}^2 - mg \cos \theta = E',$$

where E' is the total energy divided by the length of the string. Substituting the initial conditions $\dot{\theta} = \omega$ and $\theta = 0$ gives

$$\frac{1}{2}ml\omega^2 - mg = E'$$

so that

$$\frac{1}{2}ml\dot{\theta}^2 - mg \cos \theta = \frac{1}{2}ml\omega^2 - mg.$$

If the particle is to complete a vertical circle then

$$\frac{1}{2}ml\dot{\theta}^2 = mg \cos \theta + \frac{1}{2}ml\omega^2 - mg > 0$$

for all angles θ. The smallest value of the left hand side of this inequality occurs when $\cos \theta = -1$, i.e. when P is at the highest point of the circle, and so we require

$$\omega^2 > 4\frac{g}{l}.$$

This is a necessary condition for the particle to complete a vertical circle. It is not sufficient because the string could become slack before the circle is completed. Hence we also require the tension T to be non negative for all angles θ. Using the radial component of the equation of motion, together with the energy equation to eliminate $\dot{\theta}^2$, this condition can be written as

$$ml\omega^2 - 2mg + 3mg \cos \theta \geq 0,$$

for all angles θ. The smallest value of the left hand side of this inequality occurs when $\cos \theta = -1$ and so, in order that the string remains taut, we require

$$\omega^2 \geq 5\frac{g}{l}.$$

The previous inequality is satisfied if this new inequality is satisfied. The new inequality therefore becomes the necessary and sufficient condition for the completion of a vertical circle. •

TUTORIAL PROBLEM 6.4

Would the conditions on ω for the particle of Example 5 to complete a vertical circle be altered if the light inextensible string were replaced by a light rod which is free to rotate about O?

Example 6

The conical pendulum.

A conical pendulum consists of a particle P moving on a horizontal circle whilst attached to one end of a light inextensible string, the other end of which is attached to a fixed point O. Show that the angular velocity $\dot{\theta}$ of the particle in its circular path is given by

$$\dot{\theta}^2 = \frac{g}{l \cos \alpha},$$

where l is the length of the string and α is the semivertical angle of the cone generated by the string.

SOLUTION

The conical pendulum is illustrated in Fig 6.11, C being the centre of the circle on which the particle moves. Neglecting air resistance the only forces acting on the particle are its weight mg and the tension T in the string, as illustrated. Since the particle is moving on a horizontal plane the vertical component of its acceleration is zero and the vertical component of the equation of motion yields

$$0 = T \cos \alpha - mg$$

so that

$$T = \frac{mg}{\cos \alpha}.$$

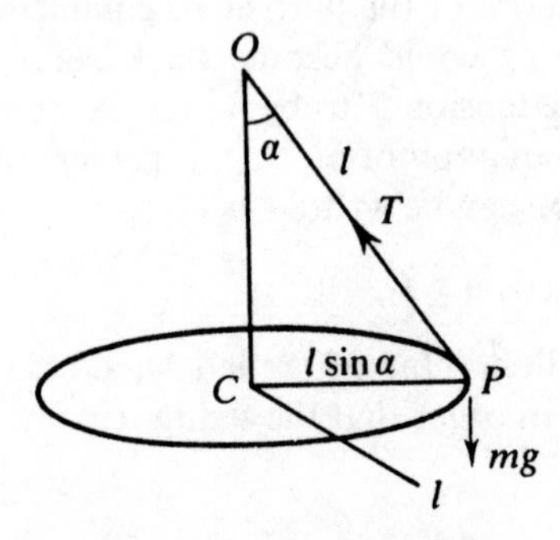

Fig 6.11 The conical pendulum.

Neither force has a component tangential to the circle and therefore the tangential component of the equation of motion gives

$$ml \sin \alpha \ddot{\theta} = 0.$$

Assuming $\sin \alpha \neq 0$ it follows that the angular velocity $\dot{\theta}$ is constant. Finally the radial component of the equation of motion becomes

$$ml \sin \alpha \dot{\theta}^2 = T \sin \alpha.$$

Assuming $\sin \alpha \neq 0$ and eliminating the tension T gives

$$\dot{\theta}^2 = \frac{g}{l \cos \alpha},$$

as required. •

In the above examples we have omitted to state what inertial frame is being used to carry out the calculations. Such an omission is common practice despite the importance of inertial frames when using Newton's second law of motion or when writing down the equation of motion. Examples 4–5 are all concerned with terrestrial motions and the fixed points, lines and planes are, although not explicitly stated, fixed relative to the earth's surface. The inertial frame $S_{\oplus}$ introduced in Section 4.2 is being used in these examples and we are assuming that the various motions are of short duration compared to the day, so that the rotation of the earth is neglected. In Example 3 we are using something of a combination of the two frames $S_{\odot}$ and $S_{\oplus}$ introduced in Section 4.2, that is an inertial frame fixed relative to the distant stars but with its origin at the centre of the earth.

TUTORIAL PROBLEM 6.5

(i) Show that the motion of a conical pendulum of given length is impossible if the magnitude of the angular velocity is less than a certain critical value.
(ii) The angular velocity of a conical pendulum of given length l and semivertical angle α has a determined constant value. Discuss how the conical pendulum might be adapted to the design of a device to govern the angular velocity of a rotating shaft.

Example 7

Rederive the results obtained in Example 2 of Chapter 4, using the fact that B and C move in circles centred at A and B respectively.

SOLUTION

B moves in a circle centred at A with angular velocity $\dot{\theta}$ and so the velocity of B relative to A has a transverse component $a\dot{\theta}$ but no radial component. C moves in a circle centred at B with angular velocity $\dot{\phi}$ and so the velocity of C relative to B has a transverse component $a\dot{\phi}$ but no radial component. These velocities are shown in Fig 6.12(i). The square of the speed of B relative to A is $a^2\dot{\theta}^2$. The velocity of C

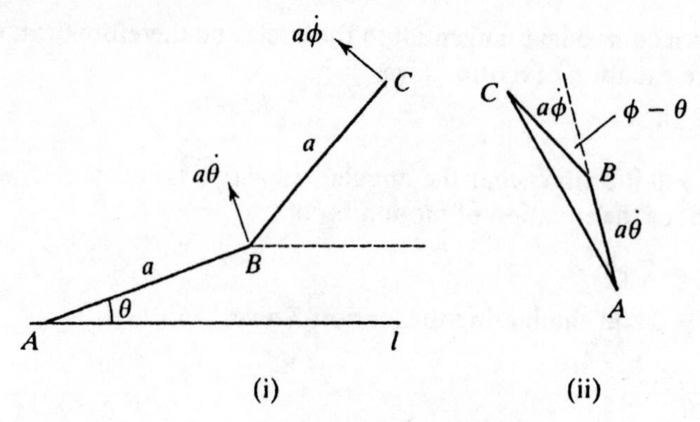

Fig 6.12 Jointed rods.

relative to A is obtained using the usual addition law of relative velocities and is represented by the third side of the vector triangle illustrated in Fig 6.12(ii). Applying the cosine rule to this triangle gives the square of the speed of C relative to A as being

$$a^2\dot{\theta}^2 + a^2\dot{\phi}^2 - 2a^2\dot{\theta}\dot{\phi}\cos(\pi - \overline{\phi - \theta})$$
$$= a^2\dot{\theta}^2 + a^2\dot{\phi}^2 + 2a^2\dot{\theta}\dot{\phi}\cos(\phi - \theta).$$

●

EXERCISES ON 6.2

1. A thin hollow tube rotates with constant angular velocity ω in a horizontal plane about a fixed point O of the tube. A particle is initially at rest in the tube at a distance d from O and the tube is smooth so that no force is acting on the particle in the radial direction. Find expressions for the velocity and displacement of the particle at any time t later.
2. A spaceship is in a circular orbit at a height of 3.40×10^5m above the earth's surface. Show that its angular velocity is 1.15×10^{-3}rad s^{-1}. What is its speed?
3. Show that a Syncom satellite cannot be used for communicating with stations in latitudes greater than 81.4°.
4. A pendulum consists of a small ball of mass m attached to a fixed hook by a long string of length l. The ball is pulled aside until the string makes an angle of 30° with the vertical, and is then released. Draw two separate diagrams indicating the forces acting on the ball

   (i) immediately after it has been released

   (ii) when it is at the lowest point of its swing. Mark clearly beside each diagram, with a broken shafted arrow (- - - ->), the direction of the acceleration at that instant.

   Calculate the magnitude of the acceleration in case (i).

5. A particle P of mass m is swinging in a fixed vertical plane at the end of a light inextensible string of length l, the other end of which is attached to a fixed point O. The particle is observed to pass its lowest point with a speed v_0. Show that

OP will become horizontal if and only if $v_0^2 \gg 2gl$. Deduce that if

$$2gl < v_0^2 < 5gl$$

the particle will fall inward from its circular path.

6. Obtain an expression for the tension in the string of the conical pendulum discussed in Example 6. Consider the limit of this tension as $\alpha \to 0$ and $\alpha \to \pi/2$.

7. The conical pendulum would seem to be a reasonable model for a fair ground ride in which suspended chairs swing out as the ride rotates. Tutorial Problem 6.5 indicates that the chairs would not swing out if the ride had not attained a certain critical angular velocity. The chairs are observed to swing out for even small angular velocities and therefore a more sophisticated model of this ride is required. Suppose that each chair has mass m and is suspended from the end of a horizontal arm of length L by means of a light rod of length l and that this rod

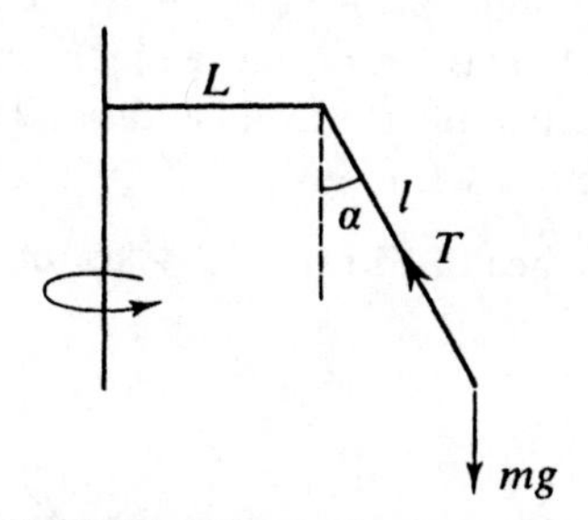

Fig 6.13 A fairground ride.

is constrained to swing in the vertical plane containing the arm as in Fig 6.13. The chair is moving on a horizontal circle when the rod connecting the chair to the arm of the ride makes an angle α with the downward vertical. Show that the arm is rotating with angular velocity $\dot{\theta}$ given by

$$\dot{\theta}^2 = \frac{g \tan \alpha}{L + l \sin \alpha}.$$

Will this result agree with the observed behaviour of the chairs?

8. A fairground ride is such that the cabin rotates in a horizontal circle of radius r_1 with constant angular velocity ω_1, the cabin itself being at the end of a long arm of radius r_2 which rotates with constant angular velocity ω_2 in a horizontal plane. At which points is the ride most thrilling?

6.3 The Use of Non-Inertial Frames of Reference

The motion of a particle P moving with constant angular velocity ω on a circle of radius r centred at O and fixed in an inertial frame S was discussed in Section 6.2, relative to an observer at O and the frame S. Let us now reconsider this motion

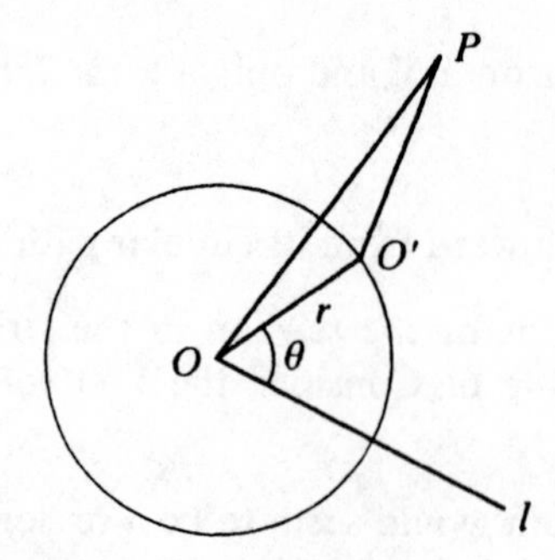

Fig 6.14 A shift of origin.

relative to a non inertial frame S' in which the particle is at rest. The simplest strategy is to shift the origin O to coincide with the location of P, we denote this new origin by O', and to choose S' to be parallel to S. This is an accelerating frame of reference. Suppose for the moment that P is a general particle of mass m and let $\mathbf{a}$ and $\mathbf{a}'$ be the accelerations of P relative to O and O', respectively. The acceleration of O' relative to O is $-r\omega^2\hat{\mathbf{r}}$. Since the frames S and S' are not rotating relative to each other we can apply the usual addition law of relative accelerations, obtained at the end of Section 4.1, to give

$$\text{acc of } P' \text{ rel to } O' = \text{acc of } P' \text{ rel to } O + \text{acc of } O \text{ rel to } O',$$

so that

$$\mathbf{a}' = \mathbf{a} + r\omega^2\hat{\mathbf{r}}.$$

The frame S is an inertial frame and therefore the equation of motion

$$m\mathbf{a} = \mathbf{F}$$

holds, where $\mathbf{F}$ is the resultant force acting on the particle P. It follows that

$$m\mathbf{a}' = \mathbf{F} + mr\omega^2\hat{\mathbf{r}}$$

and so the equation of motion of P relative to the non inertial frame S' includes an additional force $mr\omega^2\hat{\mathbf{r}}$. This force is called the **centrifugal force** and it is an example of a class of forces called **inertial forces**; these are forces which appear whenever a non inertial frame of reference is used. The magnitude of the centrifugal force is $mr\omega^2$ and it is directed away from the centre of the circle. For the original particle P, moving on the circle and located at O', the force $\mathbf{F}$ is the force of constraint which keeps the particle on its circular path and the acceleration $\mathbf{a}'$ is zero. The equation of motion of P relative to the frame S' then reduces to

$$O = \mathbf{F} + mr\omega^2\hat{\mathbf{r}}$$

so that

$$\mathbf{F} = -mr\omega^2\hat{\mathbf{r}},$$

giving the usual expression for the centripetal force.

It is very convenient to use the accelerating frame S' and the centrifugal force to solve problems. For example let us reconsider the Syncom satellite relative to the

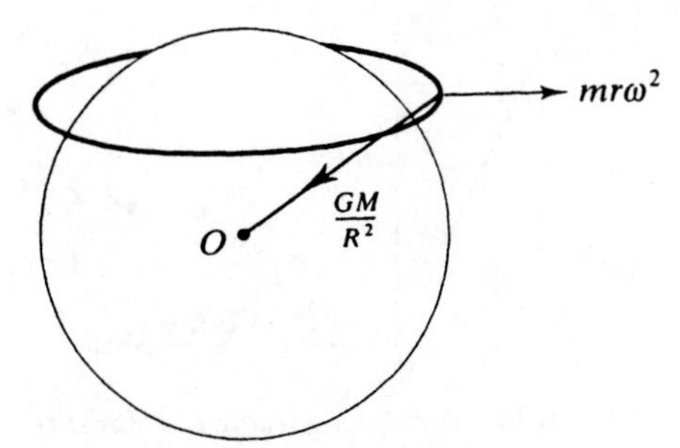

Fig 6.15 The Syncom satellite.

accelerating frame S' in which the satellite is at rest and located at the origin. The only forces acting on the satellite in this non inertial frame are the gravitational and centrifugal forces as illustrated in Fig 6.15. Here we use the same notation as in Example 3. The satellite is at rest in the frame S' and so the sum of these two forces is zero. This is only possible if the orbit lies in the equatorial plane. Then $R = r$ and

$$mr\omega^2 = \frac{GMm}{r^2}$$

from which $r = \sqrt[3]{GM/\omega^2}$.

Example 8

Explain the action of a spin drier.

SOLUTION

Once in motion the drum of a spin drier rotates with a constant angular velocity carrying wet clothing around with it. Relative to the clothing the water droplets will move under the action of the centrifugal force. They therefore move outwards through the gaps between the woven threads of the material of which the clothes are made until they come into contact with the inside of the drum. The drum is perforated and the water runs through these perforations to be pumped away. •

The frame of reference S' introduced above is useful because the particle P is at rest relative to that frame and problems concerning the motion of P relative to S are reduced to problems concerning the equilibrium of P relative to S'. Many authors use phrases like "adding the centrifugal force reduces the particle to rest". This frame S' is accelerating but not rotating relative to the inertial frame S. An alternative strategy would be to take a non inertial frame S' having the same origin as S but rotating with constant velocity ω about an axis passing through the origin and perpendicular to the plane of motion. The particle P will be at rest relative to this new frame; indeed the centrifugal force is usually associated with rotating rather than accelerating frames.

The theory of rotating frames is very important, for example it is essential to the discussion of the motion of a particle relative to the rotating earth. Traditionally this topic has not been covered in introductory courses on mechanics, but in a series of modular texts, its natural place is with the particle mechanics. The authors have felt it proper, therefore, to include an introduction here. You should not

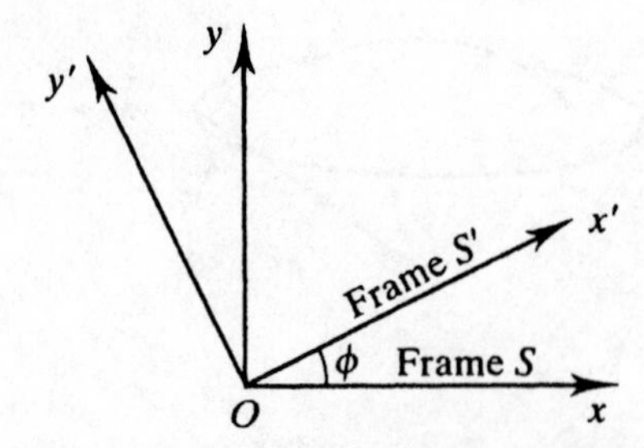

Fig 6.16 A rotating frame of reference.

worry too much if you find it difficult. Perhaps if you need to use it in later modules you will look back at the theory with added mathematical maturity and find that it is not quite so difficult after all!

Consider two frames of reference S and S' with a common origin O and suppose that S' is rotating relative to S about their common z-axis. The angle ϕ between Ox and Ox' will be a function of the time t. The cartesian basis vectors $\hat{\mathbf{i}}', \hat{\mathbf{j}}', \hat{\mathbf{k}}'$ associated with S' can be written in terms of the cartesian basis vectors $\hat{\mathbf{i}}, \hat{\mathbf{j}}, \hat{\mathbf{k}}$ associated with S as

$$\hat{\mathbf{i}}' = \hat{\mathbf{i}} \cos \phi + \hat{\mathbf{j}} \sin \phi$$

$$\hat{\mathbf{j}}' = -\hat{\mathbf{i}} \sin \phi + \hat{\mathbf{j}} \cos \phi$$

$$\text{and} \quad \hat{\mathbf{k}}' = \hat{\mathbf{k}}.$$

Now consider any vector $\mathbf{a}$. It can be written in terms of these basis vectors as either

$$\mathbf{a} = a_x\hat{\mathbf{i}} + a_y\hat{\mathbf{j}} + a_z\hat{\mathbf{k}}$$

$$\text{or} \quad \mathbf{a} = a'_x\hat{\mathbf{i}}' + a'_y\hat{\mathbf{j}}' + a_z\hat{\mathbf{k}}'.$$

From the above

$$\begin{aligned} \mathbf{a} &= a'_x\hat{\mathbf{i}}' + a'_y\hat{\mathbf{j}}' + a'_z\hat{\mathbf{k}}' \\ &= a'_x(\hat{\mathbf{i}} \cos \phi + \hat{\mathbf{j}} \sin \phi) + a'_y(-\hat{\mathbf{i}} \sin \phi + \hat{\mathbf{j}} \cos \phi) + a'_z\hat{\mathbf{k}} \\ &= (a'_x \cos \phi - a'_y \sin \phi)\hat{\mathbf{i}} + (a'_x \sin \phi + a'_y \cos \phi)\hat{\mathbf{j}} + a'_z\hat{\mathbf{k}}. \end{aligned}$$

It follows that

$$a_x = a'_x \cos \phi - a'_y \sin \phi$$

$$a_y = a'_x \sin \phi + a'_y \cos \phi$$

$$\text{and} \quad a_z = a'_z.$$

In order to distinguish between the derivatives of $\mathbf{a}$ relative to the two different frames S and S' we will use the notations

$$\frac{d\mathbf{a}}{dt} = \dot{a}_x\hat{\mathbf{i}} + \dot{a}_y\hat{\mathbf{j}} + \dot{a}_z\hat{\mathbf{k}}$$

$$\text{and} \quad \frac{D\mathbf{a}}{Dt} = \dot{a}'_x\hat{\mathbf{i}}' + \dot{a}'_y\hat{\mathbf{j}}' + \dot{a}'_z\hat{\mathbf{k}}',$$

where the dot differentiation will respect time, as usual. The relationship between these two derivatives can be found by considering $d\mathbf{a}/dt$ and eliminating the undashed components of $\mathbf{a}$ in terms of the dashed components. Thus

$$\begin{aligned}
\frac{d\mathbf{a}}{dt} &= \dot{a}_x\hat{\mathbf{i}} + \dot{a}_y\hat{\mathbf{j}} + \dot{a}_z\hat{\mathbf{k}} \\
&= \overline{(a'_x \cos\phi - a'_y \sin\phi)}^{\bullet}\hat{\mathbf{i}} + \overline{(a'_x \sin\phi + a'_y \cos\phi)}^{\bullet}\hat{\mathbf{j}} + \dot{a}'_z\hat{\mathbf{k}} \\
&= (\dot{a}'_x \cos\phi - a'_x \sin\phi\,\dot{\phi} - \dot{a}'_y \sin\phi - a'_y \cos\phi\,\dot{\phi})\hat{\mathbf{i}} + \\
&\quad (\dot{a}'_x \sin\phi + a'_x \cos\phi\,\dot{\phi} + \dot{a}'_y \cos\phi - a'_y \sin\phi\,\dot{\phi})\hat{\mathbf{j}} + \dot{a}'_z\hat{\mathbf{k}} \\
&= \dot{a}'_x(\cos\phi\,\hat{\mathbf{i}} + \sin\phi\,\hat{\mathbf{j}}) + \dot{a}'_y(-\sin\phi\,\hat{\mathbf{i}} + \cos\phi\,\hat{\mathbf{j}}) + \dot{a}'_z\hat{\mathbf{k}} \\
&\quad + a'_x\dot{\phi}(-\sin\phi\hat{\mathbf{i}} + \cos\phi\hat{\mathbf{j}}) + a'_y\dot{\phi}(-\cos\phi\hat{\mathbf{i}} - \sin\phi\hat{\mathbf{j}}) \\
&= \dot{a}'_x\hat{\mathbf{i}}' + \dot{a}'_y\hat{\mathbf{j}}' + \dot{a}'_z\hat{\mathbf{k}}' + a'_x\dot{\phi}\hat{\mathbf{j}}' - a'_y\dot{\phi}\hat{\mathbf{i}}' \\
&= \frac{D\mathbf{a}}{Dt} + \dot{\phi}(a'_x\hat{\mathbf{j}}' - a'_y\hat{\mathbf{i}}').
\end{aligned}$$

Defining the vector $\mathbf{\Omega}$ by

$$\mathbf{\Omega} = \dot{\phi}\hat{\mathbf{k}} \quad (= \dot{\phi}\hat{\mathbf{k}}')$$

it follows that

$$\begin{aligned}
\mathbf{\Omega} \times \mathbf{a} &= \dot{\phi}\hat{\mathbf{k}} \times (a'_x\hat{\mathbf{i}}' + a'_y\hat{\mathbf{j}}' + a'_z\hat{\mathbf{k}}') \\
&= \dot{\phi}(a'_x\hat{\mathbf{j}}' - a'_y\hat{\mathbf{i}}')
\end{aligned}$$

and so, finally,

$$\frac{d\mathbf{a}}{dt} = \frac{D\mathbf{a}}{Dt} + \mathbf{\Omega} \times \mathbf{a}.$$

The vector $\mathbf{\Omega}$ is called the **angular velocity vector** of S' relative to S and is directed along the axis of rotation Oz. If $\mathbf{\Omega}$ is directed in the +ve z direction then the angular velocity $\Omega = \dot{\phi}$ is positive and equal to $|\mathbf{\Omega}|$. If $\mathbf{\Omega}$ is directed in the -ve z direction then the angular velocity is negative and equal to $-|\mathbf{\Omega}|$.

The equation

$$\frac{d\mathbf{a}}{dt} = \frac{D\mathbf{a}}{Dt} + \mathbf{\Omega} \times \mathbf{a}$$

holds for any vector $\mathbf{a}$. Taking $\mathbf{a}$ to be the position vector $\mathbf{r}$ of a particle P relative to the origin O gives

$$\frac{d\mathbf{r}}{dt} = \frac{D\mathbf{r}}{Dt} + \mathbf{\Omega} \times \mathbf{r},$$

an equation which relates the velocity $d\mathbf{r}/dt$ of the particle relative to the frame S to the velocity $D\mathbf{r}/Dt$ of the particle relative to the frame S'. Taking $\mathbf{a}$ to be the derivative $d\mathbf{r}/dt$ gives

$$\frac{d^2\mathbf{r}}{dt^2} = \frac{D}{Dt}(\frac{d\mathbf{r}}{dt}) + \boldsymbol{\Omega} \times \frac{d\mathbf{r}}{dt}$$
$$= \frac{D}{Dt}(\frac{D\mathbf{r}}{Dt} + \boldsymbol{\Omega} \times \mathbf{r}) + \boldsymbol{\Omega} \times (\frac{D\mathbf{r}}{Dt} + \boldsymbol{\Omega} \times \mathbf{r}).$$

It follows that

$$\frac{d^2\mathbf{r}}{dt^2} = \frac{D^2\mathbf{r}}{Dt^2} + \frac{D\boldsymbol{\Omega}}{Dt} \times \mathbf{r} + 2\boldsymbol{\Omega} \times \frac{D\mathbf{r}}{Dt} + \boldsymbol{\Omega} \times (\boldsymbol{\Omega} \times \mathbf{r})$$

which relates the acceleration $d^2\mathbf{r}/dt^2$ of the particle relative to the frame S to the acceleration $D^2\mathbf{r}/Dt^2$ and velocity $D\mathbf{r}/Dt$ of the particle relative to the frame S'.

Now suppose that the frame of reference S is an inertial frame so that Newton's second law of motion can be used to write the equation of motion of the particle P relative to S as

$$m\frac{d^2\mathbf{r}}{dt^2} = \mathbf{F},$$

where $\mathbf{F}$ is the resultant force acting on the particle. It follows that

$$m\frac{D^2\mathbf{r}}{Dt^2} = \mathbf{F} - m\frac{D\boldsymbol{\Omega}}{Dt} \times \mathbf{r} - 2m\boldsymbol{\Omega} \times \frac{D\mathbf{r}}{Dt} - m\boldsymbol{\Omega} \times (\boldsymbol{\Omega} \times \mathbf{r})$$

so that the equation of motion of the particle P relative to the rotating, non-inertial frame S' involves three inertial forces. The second of these is called the **Coriolis force** and the third is related to the centrifugal force, as we shall see below.

Let us return to the beginning of this Section and the particle P moving with constant angular velocity ω on a circle centred at O and fixed in an inertial frame S. This particle will be at rest in the rotating frame of reference S' if the angular velocity $\Omega = \dot{\phi}$ of the frame S' relative to S is chosen to be equal to the constant angular velocity ω of the particle P relative to S. With this choice the inertial force $-m\frac{D\boldsymbol{\Omega}}{Dt} \times \mathbf{r}$ becomes zero because the angular velocity is constant. The inertial force $-2m\boldsymbol{\Omega} \times \frac{D\mathbf{r}}{Dt}$ is zero because the particle P is at rest in the rotating frame S'. The remaining inertial force $-m\boldsymbol{\Omega} \times (\boldsymbol{\Omega} \times \mathbf{r})$ can be rewritten as $-m(\boldsymbol{\Omega}.\mathbf{r}\boldsymbol{\Omega} - \boldsymbol{\Omega}.\boldsymbol{\Omega}\mathbf{r})$ which reduces to the centrifugal force $m\omega^2\mathbf{r}$ because the axis of rotation is perpendicular to the plane of motion, so that $\boldsymbol{\Omega}.\mathbf{r} = 0$, and $\Omega = \omega$.

EXERCISES ON 6.3

1. A particle is swung around on a horizontal circle with constant angular velocity ω at the end of a light inextensible string of length l. Use a rotating frame in which the particle is at rest to find the tension in the string.

2. The light inextensible string in the previous question is now replaced by a light elastic string of modulus λ and natural length l_0. Show that

$$l = \frac{\lambda l_0}{\lambda - ml_0\omega^2}.$$

What is the greatest angular velocity ω possible? Is it reasonable to assume that this value can be achieved?

3. Show that the rate of change with respect to time of the angular velocity $\mathbf{\Omega}$ is the same relative to both frames of reference S and S'.

4. Using the same notation as in the text obtain the equation

$$\frac{D^2\mathbf{r}}{Dt^2} = \frac{d^2\mathbf{r}}{dt^2} - \frac{d\mathbf{\Omega}}{dt} \times \mathbf{r} - 2\mathbf{\Omega} \times \frac{d\mathbf{r}}{dt} + \mathbf{\Omega} \times (\mathbf{\Omega} \times \mathbf{r})$$

relating the acceleration $D^2\mathbf{r}/Dt^2$ of a particle relative to the frame S' to the acceleration and velocity of the particle relative to the frame S.

5. A merry-go-round is made to rotate with a constant angular velocity ω. Discuss the effect of the Coriolis force on an attendant who attempts to walk radially outwards from the centre of the merry-go-round with constant speed u.

6.4 Motion of a Particle Relative to the Rotating Earth

The earth rotates about its axis with a constant angular velocity $\Omega = 1$ rotation per day $= 7.27 \times 10^{-5}\,\text{rad}\,\text{s}^{-1}$. This rotation has been neglected so far in our discussion of terrestrial mechanics and we have assumed that frames of reference fixed relative to the earth's surface are inertial. This has been justified by confining attention to motions of short duration compared to the day. For motions of longer duration the rotation has to be taken into account, indeed the rotation can be detected by terrestrial experiments. For example, if a simple pendulum is set swinging in a vertical plane and observed over a long period of time then its plane of motion is seen to rotate relative to the earth. Using the theory of rotating frames developed in Section 6.3 we can now investigate the effect of the earth's rotation on the motion of a particle.

The frame of reference S' with origin O at the centre of the earth and with axes fixed relative to the earth is a rotating frame. The angular velocity $\mathbf{\Omega}$ of S' relative to an inertial frame S with origin O and fixed to the distant stars has magnitude

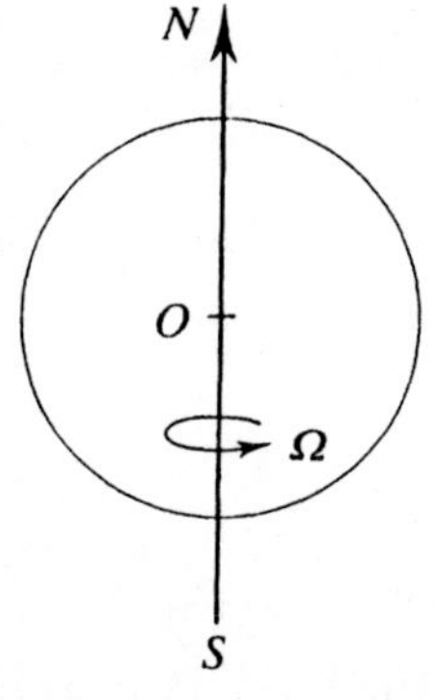

Fig 6.17 The rotating earth.

$|\mathbf{\Omega}| = 7.27 \times 10^{-5}\text{rad s}^{-1}$ and is directed along the earth's axis from south to north. Relative to this frame, the equation of motion of a given particle P of mass m moving close to the earth's surface under the action of its own weight $m\mathbf{g}$ and some other external force $\mathbf{F}_{ext}$ will take the form obtained in Section 6.3, that is

$$m\frac{D^2\mathbf{r}}{Dt^2} = m\mathbf{g} + \mathbf{F}_{ext} - 2m\mathbf{\Omega} \times \frac{D\mathbf{r}}{Dt} - m\mathbf{\Omega} \times (\mathbf{\Omega} \times \mathbf{r}).$$

Here the inertial force $-m(D\mathbf{\Omega}/Dt) \times \mathbf{r}$ has been omitted because the angular velocity vector $\mathbf{\Omega}$ is constant in magnitude and fixed in direction relative to S', so that $D\mathbf{\Omega}/Dt = 0$.

When a particle is weighed using a spring balance, the balance measures the magnitude of the external force required to maintain the particle at rest. This force is obtained from the equation of motion by putting $D\mathbf{r}/Dt = 0$ and $D^2\mathbf{r}/Dt^2 = 0$ to give

$$\begin{aligned}\mathbf{F}_{ext} &= -m\mathbf{g} + m\mathbf{\Omega} \times (\mathbf{\Omega} \times \mathbf{r}) \\ &= -m\mathbf{g}_e,\end{aligned}$$

where

$$\mathbf{g}_e = \mathbf{g} - \mathbf{\Omega} \times (\mathbf{\Omega} \times \mathbf{r}).$$

The measured weight of the particle is therefore mg_e, where $g_e = |\mathbf{g}_e|$, and it will vary from point to point on the earth's surface. The vector $\mathbf{g}_e$ is called the **effective gravitational acceleration**. For a particle P as illustrated in Fig 6.18(i) the vector product $\mathbf{\Omega} \times \mathbf{r}$ is directed into the page, that is due east. The vector triple product $\mathbf{\Omega} \times (\mathbf{\Omega} \times \mathbf{r})$ is therefore directed towards the axis of rotation. The gravitational acceleration $\mathbf{g}$ is directed towards the centre of the earth and the effective gravitational acceleration is therefore represented by the third side of the vector triangle illustrated in Fig 6.18(ii). Clearly $g_e < g$ so that the measured gravitational acceleration and weight of the particle are reduced due to the earth's rotation. Furthermore a plumbline will come to rest in the direction of $\mathbf{g}_e$ so that the observed downward vertical is no longer directed to the centre of the earth. The

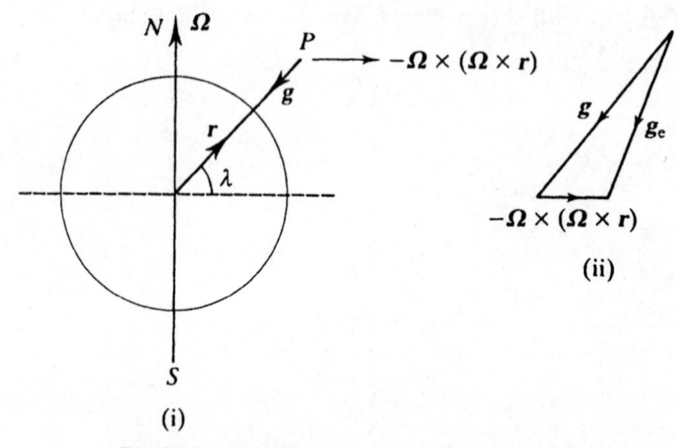

Fig 6.18 The effective gravitational acceleration.

surface of any fluid body rotating under the action of its own gravitational attraction is always perpendicular to the vertical. The fact that the vertical at different points on the surface is no longer directed towards a centre explains why the earth and other celestial bodies are oblate.

TUTORIAL PROBLEM 6.6

At what points is the effective gravitational acceleration $\mathbf{g}_e$ directed towards the centre of the earth? Can $\mathbf{g}_e$ be equal to $\mathbf{g}$?

Example 9

Prove that at a point P lying on the earth's surface

$$g_e = g(1 - \frac{\Omega^2 R \cos^2 \lambda}{g}),$$

where R is the radius of the earth, λ is the latitude of P and terms involving Ω^4 have been neglected.

SOLUTION
Referring to Fig 6.18, with P now on the earth's surface so that $r = R$,

$$\mathbf{\Omega} \times \mathbf{r} = \Omega R \sin(\frac{\pi}{2} - \lambda)\hat{\mathbf{n}} = \Omega R \cos \lambda \, \hat{\mathbf{n}},$$

where $\hat{\mathbf{n}}$ is the unit vector directed due east. Hence

$$\mathbf{\Omega} \times (\mathbf{\Omega} \times \mathbf{r}) = \mathbf{\Omega} \times \hat{\mathbf{n}}\Omega R \cos \lambda = \Omega^2 R \cos \lambda \, \hat{\mathbf{n}}',$$

where $\hat{\mathbf{n}}'$ is the unit vector directed towards the earth's axis. The effective gravitational acceleration $\mathbf{g}_e$ is the vector sum of $\mathbf{g}$ and $-\mathbf{\Omega} \times (\mathbf{\Omega} \times \mathbf{r})$ and is represented by the third side of the vector triangle in Fig 6.19. The value of g_e can be found by applying the cosine rule to this triangle. Hence

$$g_e^2 = g^2 + \Omega^4 R^2 \cos^2 \lambda - 2g\Omega^2 R \cos^2 \lambda.$$

Neglecting terms involving Ω^4 and using the Binomial Theorem gives

$$g_e = g\left(1 - \frac{2\Omega^2 R \cos^2 \lambda}{g}\right)^{1/2} = g\left(1 - \frac{\Omega^2 R \cos^2 \lambda}{g}\right).$$

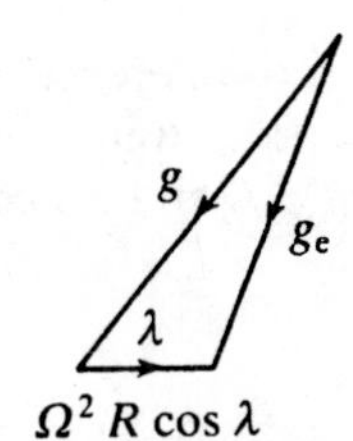

Fig 6.19 Calculating g_e.

Written in terms of the effective gravitational acceleration $\mathbf{g}_e$, the equation of motion of the particle P moving close to the earth's surface becomes

$$m\frac{D^2\mathbf{r}}{Dt^2} = m\mathbf{g}_e + \mathbf{F}_{\text{ext}} - 2m\mathbf{\Omega} \times \frac{D\mathbf{r}}{Dt}.$$

In what follows $\mathbf{g}_e$ will be taken to be a constant vector with the magnitude found in Example 9 and in the direction of the local downward vertical in the vicinity of the motion. Let us now return to the simple pendulum and the particle moving on a vertical circle as discussed in Examples 4 and 5. Both of the forces acting on the particle in these examples, that is the tension and weight, lie in the fixed vertical plane of motion. Taking into account the rotation of the earth a third force has to be considered, namely the Coriolis force $-2m\mathbf{\Omega} \times D\mathbf{r}/Dt$. This force is perpendicular to the earth's axis and so parallel to the equatorial plane. It follows that if the plane of motion is not parallel to the equatorial plane then the Coriolis force will have a component perpendicular to the plane of motion which cannot, therefore, be fixed relative to the rotating earth. This is the principle behind the Foucault pendulum, a very long and heavy pendulum which can continue to swing over a long enough period of time to enable an observer to detect a rotation of the plane of motion with angular velocity $\Omega \sin \lambda$. In the Northern hemisphere this rotation is clockwise, looking from above the pendulum, and in the Southern hemisphere it is anticlockwise.

You will have noticed that we have not actually derived the expression for the angular velocity of rotation of the plane of motion of the simple pendulum, quoted above. The derivation is somewhat complicated. However in the next example we return to Example 4 of Chapter 4 and actually solve the equation of motion in order to find the deviation of a projectile from its plane of projection due to the presence of the Coriolis force.

Example 10

The easterly drift.

A projectile P is fired from a point O on the earth's surface with velocity $\mathbf{u}$ relative to O and a frame of reference rotating with the earth. Find an expression for the position vector $\mathbf{r}$ of P relative to O at a subsequent time t, air resistance being neglected. Show that at time t the projectile has deviated from the plane of projection by an amount $\frac{1}{2}\Omega g_e \cos \lambda t^3$, in the easterly direction.

SOLUTION

Translating the origin of the frame of reference S' rotating with the earth from the centre of the earth to the point of projection does not alter the equation of motion of any given particle. Putting $\mathbf{F}_{\text{ext}} = 0$ and cancelling the mass m, the equation of motion of the projectile reduces to

$$\frac{D^2\mathbf{r}}{Dt^2} = \mathbf{g}_e - 2\mathbf{\Omega} \times \frac{D\mathbf{r}}{Dt}.$$

This equation can be integrated to give

$$\frac{D\mathbf{r}}{Dt} = \mathbf{g}_e t - 2\boldsymbol{\Omega} \times \mathbf{r} + \mathbf{k}_1,$$

where $\mathbf{k}_1$ is a constant vector of integration and it has been assumed that $\mathbf{g}_e$ is constant. A second integration cannot be carried out directly. However this equation can be used to eliminate $D\mathbf{r}/Dt$ from the previous equation, yielding

$$\frac{D^2\mathbf{r}}{Dt^2} = \mathbf{g}_e - 2\boldsymbol{\Omega} \times \mathbf{g}_e t + 4\boldsymbol{\Omega} \times (\boldsymbol{\Omega} \times \mathbf{r}) - 2\boldsymbol{\Omega} \times \mathbf{k}_1.$$

We now appeal to the size of the angular velocity of rotation of the earth, $7.27 \times 10^{-5}\,\text{rad s}^{-1}$, to justify neglecting the term $4\boldsymbol{\Omega} \times (\boldsymbol{\Omega} \times \mathbf{r})$ which involves the square of this small quantity. Repeated integration of the resulting equation gives

$$\frac{D\mathbf{r}}{Dt} = \mathbf{g}_e t - \boldsymbol{\Omega} \times \mathbf{g}_e t^2 - \boldsymbol{\Omega} \times \mathbf{k}_1 t + \mathbf{k}_2$$

and

$$\mathbf{r} = \frac{1}{2}\mathbf{g}_e t^2 - \frac{1}{3}\boldsymbol{\Omega} \times \mathbf{g}_e t^3 - \boldsymbol{\Omega} \times \mathbf{k}_1 t^2 + \mathbf{k}_2 t + \mathbf{k}_3.$$

By a somewhat round about route we have obtained an expression for the position vector of the projectile relative to the point of projection as a function of time involving three arbitrary constant vectors of integration, that is one too many. Substituting the expression for $\mathbf{r}$ into the original equation

$$\frac{D^2\mathbf{r}}{Dt^2} = \mathbf{g}_e - 2\boldsymbol{\Omega} \times \frac{D\mathbf{r}}{Dt}$$

and neglecting terms involving squares of Ω give

$$\mathbf{g}_e - 2\boldsymbol{\Omega} \times \mathbf{g}_e t - 2\boldsymbol{\Omega} \times \mathbf{k}_1 = \mathbf{g}_e - 2\boldsymbol{\Omega} \times (\mathbf{g}_e t + \mathbf{k}_2)$$

$$\text{or} \qquad \boldsymbol{\Omega} \times \mathbf{k}_1 = \boldsymbol{\Omega} \times \mathbf{k}_2.$$

This can be used to eliminate $\boldsymbol{\Omega} \times \mathbf{k}_1$ from the expression for $\mathbf{r}$, leaving

$$\mathbf{r} = \frac{1}{2}\mathbf{g}_e t^2 - \frac{1}{3}\boldsymbol{\Omega} \times \mathbf{g}_e t^3 - \boldsymbol{\Omega} \times \mathbf{k}_2 t^2 + \mathbf{k}_2 t + \mathbf{k}_3.$$

Putting $\mathbf{k}_2 = \mathbf{c}_1$, $\mathbf{k}_3 = \mathbf{c}_2$ and rearranging yields

$$\mathbf{r} = \frac{1}{2}\mathbf{g}_e t^2 + \mathbf{c}_1 t + \mathbf{c}_2 - \frac{1}{3}\boldsymbol{\Omega} \times \mathbf{g}_e t^3 - \boldsymbol{\Omega} \times \mathbf{c}_1 t^2.$$

You will recognize the first three terms from the solution to Example 4 of Chapter 4. Since the projectile is projected from the origin O at time $t = 0$ it follows that $\mathbf{c}_2 = 0$ and that c_1 is the velocity of projection $\mathbf{u}$, so that

$$\mathbf{r} = \frac{1}{2}\mathbf{g}_e t^2 + \mathbf{u}t - \frac{1}{3}\boldsymbol{\Omega} \times \mathbf{g}_e t^3 - \boldsymbol{\Omega} \times \mathbf{u}t^2.$$

The last two terms are "correction" terms to be added to the result in Chapter 4 in order to account for the rotation of the earth. The second of these correction terms can be combined with the term $\frac{1}{2}\mathbf{g}_e t^2$ so that, without the term $-\frac{1}{3}\boldsymbol{\Omega} \times \mathbf{g}_e t^3$ the path of the projectile lies in the plane passing through the point of projection and containing the vectors $(\mathbf{g}_e - 2\mathbf{\Omega} \times \mathbf{u})$ and $\mathbf{u}$. We shall call this plane the plane of projection, it is no longer a vertical plane. The term $-\frac{1}{3}\mathbf{\Omega} \times \mathbf{g}_e t^3$ then represents a deviation from this plane. We can see from Fig 6.17 that this deviation is directed due east and has magnitude $\frac{1}{3}\Omega g_e \cos \lambda\, t^3$. •

TUTORIAL PROBLEM 6.7

In the solution to the last example it was stated that translating the origin of the frame of reference rotating with the earth from the centre of the earth to the point of projection does not alter the equation of motion. Discuss the justification for this statement.

EXERCISES ON 6.4

1. Taking $R = 6.36 \times 10^6$m, calculate the greatest percentage decrease in the gravitational acceleration due to the rotation of the earth.

2. Using Fig 6.19 show that the local vertical at a point P lying on the earth's surface deviates from the line joining P to the centre of the earth by an angle δ given by

$$\delta = \frac{\Omega^2 R \sin 2\lambda}{2g},$$

terms involving Ω^4 being neglected. What is the greatest numerical value of this deviation?

3. Will the conical pendulum of Example 6 still move with constant angular velocity when the Coriolis force is taken into account?

4. A particle is released from rest from the top of a tower of height h at latitude λ. Show that the particle will hit the ground a distance $\frac{2}{3}\Omega h\sqrt{2h/g}\cos \lambda$ east of the tower, air resistance being neglected.

Summary

- using **plane polar coordinates** the velocity and acceleration of a particle are given by

$$\mathbf{v} = \dot{r}\hat{\mathbf{r}} + r\dot{\theta}\hat{\boldsymbol{\theta}}$$

and $$\mathbf{a} = (\ddot{r} - r\dot{\theta}^2)\hat{\mathbf{r}} + (2\dot{r}\dot{\theta} + r\ddot{\theta})\hat{\boldsymbol{\theta}}$$

- the components of a vector relative to the basis vectors $\hat{\mathbf{r}}$ and $\hat{\boldsymbol{\theta}}$ are called the **radial** and **transverse** components, respectively

- the **centripetal force** is the force which constrains a particle to move on a circle with constant angular velocity relative to an inertial frame; it has magnitude $ma\omega^2$ and is directed towards the centre of the circle

- the period of a **simple pendulum** of length l is $2\pi\sqrt{l/g}$

- the **centrifugal force** is the inertial force which is exerted on a particle moving on a circle with constant angular velocity relative to an inertial frame when viewed from a non inertial frame in which the particle is at rest; it has magnitude $ma\omega^2$ and is directed away from the centre of the circle

- if $\mathbf{\Omega}$ is the **angular velocity vector** of a frame S' relative to S then

$$\frac{d\mathbf{a}}{dt} = \frac{D\mathbf{a}}{Dt} + \mathbf{\Omega} \times \mathbf{a},$$

  where d/dt and D/Dt denote the derivatives relative to S and S' respectively

- the equation of motion of a particle relative to the rotating earth is

$$m\frac{D^2\mathbf{r}}{Dt^2} = m\mathbf{g}_e + \mathbf{F}_{\text{ext}} - 2m\mathbf{\Omega} \times \frac{D\mathbf{r}}{Dt}$$

  where $\mathbf{g}_e$ is the **effective gravitational acceleration** defined by $\mathbf{g}_e = \mathbf{g} - \mathbf{\Omega} \times (\mathbf{\Omega} \times \mathbf{r})$ and $-2m\mathbf{\Omega} \times D\mathbf{r}/Dt$ is the Coriolis force.

FURTHER EXERCISES

1. The leminscate of Bernoulli is the curve with polar equation

$$r^2 = a^2 \cos 2\theta.$$

A particle moves on such a curve with a speed v. Show that

$$\dot{r}^2 = \frac{v^2}{a^4}(a^4 - r^4).$$

2. A bead can slide freely on a straight wire AB of length l, which is rotated in a horizontal plane with constant angular velocity ω about the fixed end A. Initially the bead is projected along the wire with speed u from the end A. Show that, if r is the distance travelled along the wire in time t, then

$$r = \frac{u}{2\omega}(e^{\omega t} - e^{-\omega t})$$

and that the bead arrives at the end B of the wire at time

$$t = \frac{1}{\omega}\sinh^{-1}\left(\frac{\omega l}{u}\right).$$

Show also that the radial velocity is then $(u^2 + \omega^2 l^2)^{\frac{1}{2}}$.

3. A particle P describes a circle with constant angular velocity ω. Show that the perpendicular projection of P onto a diameter performs simple harmonic motion.

4. A solid cone of semivertical angle α is rotating with constant angular velocity ω about its axis which is vertical. A particle P of mass m is suspended from the fixed vertex O at the end of a light inextensible string of length l_0. The tension T in the string, weight mg of the particle and normal reaction R of the cone on the particle are illustrated in Fig 6.20, it being assumed that the frictional force acting on the particle is sufficiently large to prevent the particle from slipping on the cone so that the particle is carried around by the cone. Show that the particle

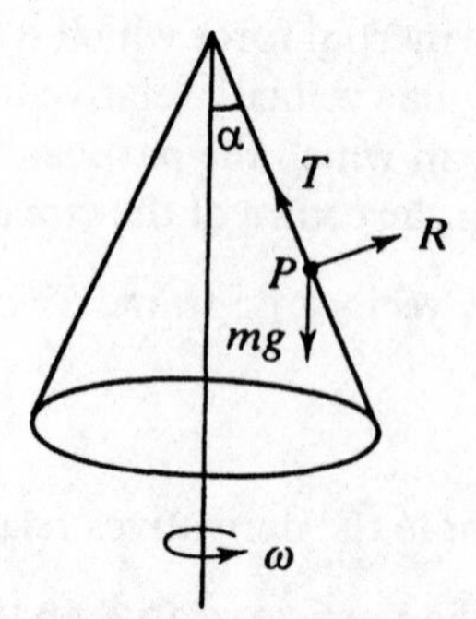

Fig 6.20 Particle on a rotating cone.

will remain in contact with the cone provided that $\omega^2 \leq g/l\cos\alpha$. What will happen if the angular velocity is slowly increased beyond the value $\sqrt{g/l\cos\alpha}$?

5. You and a friend are stood on a merry-go-round on opposite sides of the centre. The merry-go-round rotates with constant angular velocity. What happens if you try to throw a ball to each other?

6. The Coriolis force is responsible for the tradewinds. Discuss.

7 • Central Forces

The path of a particle moving under a central force must lie on a plane, a result which follows from the conservation of angular momentum. Inverse square central forces are unique in so far as for such central forces alone a second conserved vector exists, the Lenz-Runge vector. This general theory is used to derive the polar equation of the orbit of a particle moving under the gravitational attraction of a second particle. This polar equation is analysed without any assumption of previous knowledge of the conics which it represents. The orbiting particle is shown to model many of the motions which occur in celestial mechanics. In particular this model leads to a derivation of Kepler's three laws of planetary motion.

7.1 General Results

A force $\mathbf{F}$ acting on a given particle P is called a **central force** if and only if it is directed towards or away from a given origin O and its magnitude is a function of the distance of P from that origin. For such a force

$$\mathbf{F} = F(r)\hat{\mathbf{r}}$$

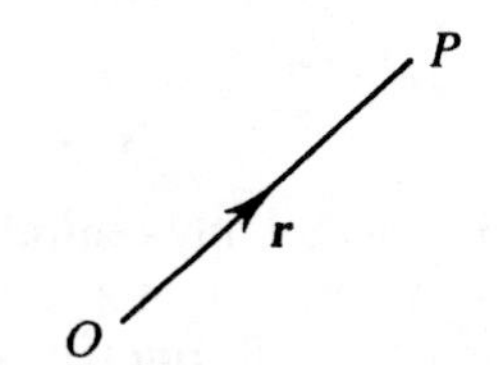

Fig 7.1 Particle and origin.

where $F(r)$, the radial component of the force, can take positive and negative values. When $F(r)$ is positive the force is directed away from the origin and is said to be **repulsive**. When $F(r)$ is negative the force is directed towards the origin and is said to be **attractive**. If the point O is fixed in an inertial frame then the equation of motion of a particle of mass m moving under the action of the central force is

$$m\ddot{\mathbf{r}} = F(r)\hat{\mathbf{r}} = \frac{F(r)}{r}\mathbf{r}.$$

The moment $\mathbf{M}$ of the force about the origin O is

$$\mathbf{M} = \mathbf{r} \times \mathbf{F} = \mathbf{r} \times \left(\frac{F(r)}{r}\mathbf{r}\right) = \frac{F(r)}{r}\mathbf{r} \times \mathbf{r} = 0,$$

a result which also follows trivially from the fact that the line of action of the force always passes through the origin. It follows that the angular momentum of the particle is conserved so that the vector

$$\mathbf{L} = m\mathbf{r} \times \dot{\mathbf{r}}$$

remains constant throughout the motion of the particle. The path of the particle P therefore lies on a fixed plane passing through O and perpendicular to $\mathbf{L}$, see Example 6 of Section 4.4. This plane will be denoted by π. The motion of the particle in the plane π is determined by those components of the equation of motion lying in π. That component of the equation of motion which is perpendicular to π can be eliminated by taking the vector product of the equation with $\mathbf{L}$, a result which follows from the fact that the normal to the plane is in the direction of $\mathbf{L}$ and $\mathbf{L} \times \mathbf{L} = 0$. We call the resulting vector equation,

$$m\ddot{\mathbf{r}} \times \mathbf{L} = \frac{F(r)}{r}\mathbf{r} \times \mathbf{L},$$

the modified equation of motion. It can be written as

$$\frac{d}{dt}(m\dot{\mathbf{r}} \times \mathbf{L}) = m\frac{F(r)}{r}\mathbf{r} \times (\mathbf{r} \times \dot{\mathbf{r}})$$

or, cancelling the mass m and using the well known identity for the vector triple product,

$$\frac{d}{dt}(\dot{\mathbf{r}} \times \mathbf{L}) = \frac{F(r)}{r}[(\mathbf{r}.\dot{\mathbf{r}})\mathbf{r} - (\mathbf{r}.\mathbf{r})\dot{\mathbf{r}}].$$

Differentiating $\mathbf{r}.\mathbf{r} = r^2$ gives $\mathbf{r}.\dot{\mathbf{r}} = r\dot{r}$ and so the above equation can be rewritten as

$$\frac{d}{dt}(\dot{\mathbf{r}} \times \mathbf{L}) = F(r)\dot{r}\mathbf{r} - F(r)r\dot{\mathbf{r}}.$$

The intriguing question to ask is whether any central forces $\mathbf{F}$ exist for which the right hand side of this equation can be written as a derivative, thus enabling the equation to be integrated trivially? If the right hand side can be written as a derivative then that derivative must be of the form

$$\frac{d}{dt}(\alpha(r)\mathbf{r}) = \frac{d\alpha}{dt}\mathbf{r} + \alpha\frac{d\mathbf{r}}{dt}.$$

Thus, in order to integrate the equation trivially, the central force must be such that a function $\alpha(r)$ exists with

$$\frac{d\alpha}{dt} = F(r)\frac{dr}{dt} \quad \text{and} \quad \alpha = -F(r)r.$$

Eliminating $F(r)$ between these equations yields

$$\frac{d\alpha}{dt} = -\frac{\alpha}{r}\frac{dr}{dt}$$

or

$$r\frac{d\alpha}{dt} + \alpha\frac{dr}{dt} = \frac{d}{dt}(r\alpha) = 0,$$

so that

$$\alpha = \frac{\lambda}{r},$$

where λ is an arbitrary constant. Substituting this into the equation $\alpha = -F(r)r$ gives

$$F(r) = -\frac{\lambda}{r^2}.$$

The search for mathematical simplicity, that is the possibility of integrating the modified equation of motion trivially, has led uniquely to the most important of all the central forces, namely the **inverse square force**. With $\lambda = Gmm'$ we get the Newtonian gravitational force of attraction between two particles, discussed in Section 5.1, and with $\lambda = -qq'/4\pi\epsilon$ we get the Coulomb electrostatic force between two static particles carrying charges q and q', ϵ being a constant called the permittivity.

For an inverse square force the modified equation of motion becomes simply

$$\frac{d}{dt}(\dot{\mathbf{r}} \times \mathbf{L}) = \frac{d}{dt}(\alpha(r)\mathbf{r}) = \frac{d}{dt}(\lambda\frac{\mathbf{r}}{r}).$$

Integrating this equation yields

$$\dot{\mathbf{r}} \times \mathbf{L} - \lambda\frac{\mathbf{r}}{r} = \mathbf{k},$$

where $\mathbf{k}$ is a constant vector of integration. This vector is called the Lenz-Runge vector; it appeared in a paper by W.Lenz published in 1924 and in one of the first text books on vectors, written by C.Runge and published in 1919. However it was discovered earlier, in 1845, by the Irish mathematician Sir William Hamilton (1805–1865). Vectors were unknown in 1845 but Hamilton had introduced a mathematical structure known as the theory of quaternions. Quaternions are related to three dimensional vectors in much the same way as complex numbers are related to two dimensional vectors. The Astronomer Royal of the day, Sir William Herschel, advised Hamilton to apply his theory of quaternions to the study of planetary motion. Doing so, Hamilton was able to give a very simple derivation of the fact that the motion is planar and also found the analogue of the Lenz-Runge vector. The history of this vector goes even further back, before the advent of either vectors or quaternions, to Laplace who found three scalars which remain constant throughout the motion of a planet. These scalars are the components of the Lenz-Runge vector.

TUTORIAL PROBLEM 7.1

Show that the Lenz-Runge vector lies in the plane of motion and deduce that $\mathbf{L}, \mathbf{k}$ and $\mathbf{L} \times \mathbf{k}$ form an orthogonal set of basis vectors at each end point of the path of a particle moving under an inverse square central force.

The energy equation for a central force was found in Example 7 of Section 4.4. It can be written in the form

$$\frac{1}{2}m|\dot{\mathbf{r}}|^2 - \int F(r)dr = E.$$

For the inverse square force $F(r) = -\lambda/r^2$ the potential is

$$V(r) = -\int F(r)dr = \int \frac{\lambda}{r^2}dr = -\frac{\lambda}{r} + \text{constant}.$$

It is conventional to take this potential to be zero at infinity. Then

$$V(r) = -\frac{\lambda}{r}$$

and

$$\frac{1}{2}m\dot{\mathbf{r}}.\dot{\mathbf{r}} - \frac{\lambda}{r} = E.$$

The magnitude of the Lenze-Runge vector is related to the total energy E. To see this consider

$$\begin{aligned}|\mathbf{k}|^2 = \mathbf{k}.\mathbf{k} &= (\dot{\mathbf{r}} \times \mathbf{L} - \lambda\frac{\mathbf{r}}{r}).(\dot{\mathbf{r}} \times \mathbf{L} - \lambda\frac{\mathbf{r}}{r}) \\ &= (\dot{\mathbf{r}} \times \mathbf{L}).(\dot{\mathbf{r}} \times \mathbf{L}) - 2\lambda\frac{\mathbf{r}}{r}.(\mathbf{r} \times \mathbf{L}) + \lambda^2\frac{\mathbf{r}.\mathbf{r}}{r^2}.\end{aligned}$$

From $\sin^2\theta = 1 - \cos^2\theta$ it follows that

$$(\dot{\mathbf{r}} \times \mathbf{L}).(\dot{\mathbf{r}} \times \mathbf{L}) = |\dot{\mathbf{r}}|^2L^2 - (\dot{\mathbf{r}}.\mathbf{L})^2 = |\dot{\mathbf{r}}|^2L^2,$$

since $\dot{\mathbf{r}}.\mathbf{L} = 0$. Also

$$\mathbf{r}.(\dot{\mathbf{r}} \times \mathbf{L}) = \mathbf{L}.(\mathbf{r} \times \dot{\mathbf{r}}) = \mathbf{L}.(\mathbf{L}/m) = L^2/m.$$

Hence

$$\begin{aligned}|\mathbf{k}|^2 &= |\dot{\mathbf{r}}|^2L^2 - 2\frac{\lambda L^2}{mr} + \lambda^2 \\ &= \frac{2L^2}{m}(E + \frac{\lambda}{r}) - \frac{2\lambda L^2}{mr} + \lambda^2 \\ &= \frac{2L^2E}{m} + \lambda^2.\end{aligned}$$

For the Newtonian gravitational force $\lambda = Gmm'$ and then

$$|\mathbf{k}|^2 = \frac{2L^2E}{m} + G^2m^2m'^2.$$

EXERCISES ON 7.1

1. A particle of mass m moves under the action of a central force with $F(r) = -\mu r$, where $\mu > 0$. Show that

$$\mathbf{r} = \mathbf{a}\sin\,\omega t + \mathbf{b}\cos\,\omega t$$

satisfies the equation of motion, where $\omega^2 = \mu/m$, and **a** and **b** are two arbitrary constant vectors. Find expressions for the angular momentum **L** and total energy E of the particle in terms of $\mathbf{a}, \mathbf{b}, m$ and ω.

2. A particle of mass m moves under the action of a central force $-\lambda\mathbf{r}/r^3$. Show directly that the derivatives with respect to time of the two vectors

$$\mathbf{L} = m\mathbf{r} \times \dot{\mathbf{r}} \quad \text{and} \quad \mathbf{k} = \dot{\mathbf{r}} \times \mathbf{L} - \lambda\frac{\mathbf{r}}{r}$$

are both zero. State the consequences of this result.

7.2 Orbit Theory

In this section we consider the motion of two spherically symmetric bodies, each moving under the gravitational attraction of the other, under the assumption that one of the bodies is at rest in an inertial frame. The second body can then be thought of as orbiting about the first. You will learn in Section 7.4 that this assumption is equivalent to the assumption that the mass of the first body is very much larger than the mass of the second body and will find there a discussion of the more general situation when this assumption is not made.

If the radius of the orbiting body is small compared to the distance between the two bodies then that body can be modelled as a particle. This particle will be denoted by P. Using the result found in Section 5.2 the gravitational attraction between P and the fixed body can be calculated as if the whole mass of the fixed body were concentrated at its centre. Hence the fixed body can also be modelled as a particle. This particle will be denoted by O and the motion of the two bodies is now modelled by the motion of a particle P orbiting about a fixed particle O. The same model can be used if the orbiting body is not spherically symmetric provided that its largest linear dimension is small compared to the distance between the bodies.

If the radius of the spherically symmetric orbiting body is not negligible the force which the fixed body exerts on each element of the orbiting body can be calculated as if the mass of the fixed body were concentrated at its centre. The total force exerted on the orbiting body can then be obtained by integration. The integration is analogous to that in Section 5.2 and it follows that the total force can be calculated as if the mass of the orbiting body were concentrated at its centre. Thus, so far as the total gravitational force is concerned, both bodies can be considered as particles and the model introduced in the last paragraph can be used to discuss the motion of the centre of the orbiting body relative to the centre of the fixed body.

The above particle model is valid in the case of the motion of a planet about the sun or the motion of the moon or of an artificial satellite about the earth, the oblateness of the sun, earth and other planets being neglected. In each case the force which the particle O exerts on P is an inverse square central force with $\lambda = Gmm'$, where m is the mass of P and m' is the mass of O. The path of the particle P is called its **orbit** and, according to the general result in Section 7.1, this orbit lies on a plane π containing O. In terms of plane polar coordinates defined in π with O as origin,

$$\mathbf{r} = r\hat{\mathbf{r}} \quad \text{and} \quad \dot{\mathbf{r}} = \dot{r}\hat{\mathbf{r}} + r\dot{\theta}\hat{\boldsymbol{\theta}}$$

so that

$$\mathbf{r} \times \dot{\mathbf{r}} = r^2\dot{\theta}\hat{\mathbf{n}} \quad \text{and} \quad \dot{\mathbf{r}}.\dot{\mathbf{r}} = \dot{r}^2 + r^2\dot{\theta}^2,$$

where $\hat{\mathbf{n}}$ is the unit vector $\hat{\mathbf{r}} \times \hat{\boldsymbol{\theta}}$, perpendicular to π. Substituting these expressions into the equations

$$m\mathbf{r} \times \dot{\mathbf{r}} = \mathbf{L}$$

and

$$\frac{1}{2}m\dot{\mathbf{r}}.\dot{\mathbf{r}} - \frac{Gmm'}{r} = E,$$

which express the conservation of angular momentum and of energy, yields

$$mr^2\dot{\theta} = L$$

and

$$\frac{1}{2}m(\dot{r}^2 + r^2\dot{\theta}^2) - \frac{Gmm'}{r} = E,$$

respectively. Here L is the component of the angular momentum normal to the plane of the orbit. Eliminating $\dot{\theta}$ between these two equations gives

$$\frac{1}{2}m\dot{r}^2 + U(r) = E,$$

$$\text{where} \quad U(r) = \frac{L^2}{2mr^2} - \frac{Gmm'}{r}.$$

This equation is analogous to the energy equation for one dimensional motion under a potential $U(r)$, indeed $U(r)$ is called the **effective potential.**

The motion of P can be discussed qualitatively by methods analogous to those introduced in Section 2.5. The graph of $U(r)$ against r is sketched in Fig 7.2. The minimum value of $U(r)$ is $-Gm^3m'^2/2L^2$ and occurs at $r = L^2/Gm^2m'$. Since $\dot{r}^2 \geq 0$, motion is only possible at points whose coordinate r is such that $U(r) \leq E$. If $E = -Gm^3m'^2/2L^2$ motion can only take place with $r = L^2/Gm^2m'$ so that the particle P moves on a circular orbit. This would correspond, for example, to a Syncom satellite. Notice that if this orbit is slightly disturbed then r will oscillate between two values close to $r = L^2/Gm^2m'$. In other words the circular orbit is stable in the sense that a slight disturbance will not cause the orbit to deviate far from the undisturbed circular orbit. If $0 > E > -Gm^3m'^2/2L^2$ then r will oscillate between the values r_1 and r_2 as illustrated in Fig 7.2. The particle P therefore moves on a closed orbit which is typical of the planets. The points of a planetary orbit corresponding to $r = r_1$, the closest approach to the sun, and $r = r_2$, the furthest retreat from the sun, are called **perihelion** and **aphelion**, respectively. For a satellite orbiting the earth these points are called **perigee** and **apogee**, respectively. Notice that at these points $E = U(r)$ so that $\dot{r} = 0$ and the radial velocity becomes zero. Any point at which the radial velocity vanishes is called an **apse**. Finally, if $E \geq 0$

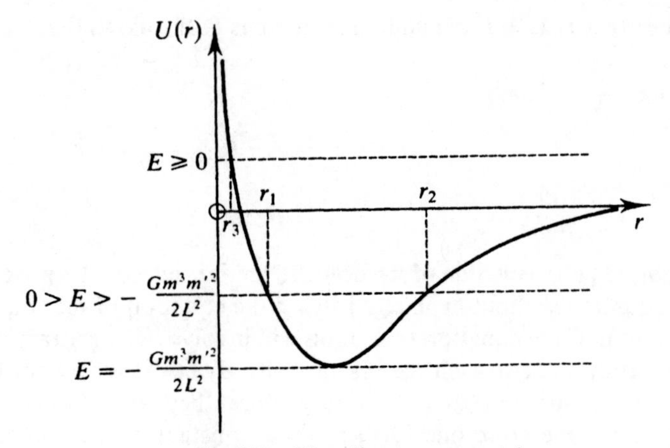

Fig 7.2 The effective potential.

then r will tend to infinity, perhaps after having first decreased to the minimum value r_3, as illustrated in Fig 7.2. The particle P therefore moves on an open orbit which is typical of certain comets; those comets which return periodically to the solar system must move on closed orbits and so correspond to $0 > E > -Gm^3m'^2/2L^2$. These results are summarized in Table 7.1. The polar

Conditions On E	Type of orbit
$E = -Gm^3m'^2/2L^2$	circular
$0 > E > -Gm^3m'^2/2L^2$	closed
$E \geq 0$	open

Table 7.1 Types of orbit

equation of the orbit of P can be found directly using the Lenz-Runge vector $\mathbf{k}$. As was seen in the last section $\mathbf{k}$ is a constant vector lying in the plane of the orbit, see Tutorial Problem 7.1. Consider plane polar coordinates with origin at O and initial line $\theta = 0$ in the direction of $\mathbf{k}$. Then

$$\begin{aligned}\mathbf{k}.\mathbf{r} &= (\dot{\mathbf{r}} \times \mathbf{L} - Gmm'\frac{\mathbf{r}}{r}).\mathbf{r} \\ &= (\dot{\mathbf{r}} \times \mathbf{L}).\mathbf{r} - Gmm'r.\end{aligned}$$

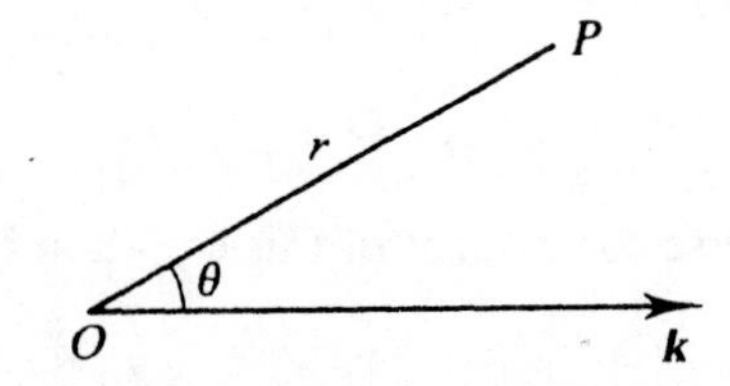

Fig 7.3 Choice of polar coordinates.

Now $(\dot{\mathbf{r}} \times \mathbf{L}).\mathbf{r} = (\mathbf{r} \times \dot{\mathbf{r}}).\mathbf{L} = L^2/m$ and $\mathbf{k}.\mathbf{r} = |\mathbf{k}|r \cos\theta$. It follows that

$$|\mathbf{k}|r \cos\theta = \frac{L^2}{m} - Gmm'r$$

or

$$\frac{1}{r} = \frac{Gmm'}{L^2}(1 + \frac{|\mathbf{k}|}{Gmm'}\cos\theta).$$

This is the required **polar equation of the orbit**. It should cause no surprise to have obtained this equation without explicitly solving the equation of motion because the general solution of the equation of motion will involve two arbitrary constant vectors of integration and two such vectors have already been found, namely **L** and **k**. In fact **L** and **k** are not quite arbitrary since they are always mutually orthogonal. There is therefore one further scalar constant of integration in the complete solution of the equation of motion and this arises when $mr^2\dot{\theta} = L$ is integrated to find θ as a function of time.

The polar equation of the orbit is usually written as

$$\frac{1}{r} = \frac{1}{l}(1 + e\cos\theta),$$

where

$$l = \frac{L^2}{Gm^2m'}$$

and

$$e = \frac{|\mathbf{k}|}{Gmm'}.$$

The constant l is called the **semilatus rectum** of the orbit and e is called the **eccentricity**. Notice that here $e \geq 0$. The relationship between the magnitude $|\mathbf{k}|$ of the Lenz-Runge vector and the total energy E was obtained at the end of Section 7.1. Using that result we can rewrite

$$e = \left(\frac{2L^2E}{G^2m^3m'^2} + 1\right)^{1/2}.$$

TUTORIAL PROBLEM 7.2

The equation of motion of a particle P of mass m orbiting about a fixed particle O of mass m' under the gravitational attraction $-Gmm'/r^2$ can be written in polar coordinates as

$$m[(\ddot{r} - r\dot{\theta}^2)\hat{\mathbf{r}} + (2\dot{r}\dot{\theta} + r\ddot{\theta})\hat{\theta}] = -\frac{Gmm'}{r^2}\hat{\mathbf{r}}.$$

Prove that the transverse component of this equation can be integrated to yield

$$mr^2\dot{\theta} = L,$$

where L is a constant of integration. Using this, show that if $r = 1/u$, then

$$\frac{dr}{dt} = -\frac{L}{m}\frac{du}{d\theta} \quad \text{and} \quad \frac{d^2r}{dt^2} = -\frac{L^2u^2}{m^2}\frac{d^2u}{d\theta^2}$$

and deduce that the radial component of the equation of motion reduces to a simple harmonic motion equation when rewritten in terms of the function $u(\theta)$. Hence rederive the polar equation of the orbit obtained in the text.

The polar equation

$$\frac{1}{r} = \frac{1}{l}(1 + e\cos\theta)$$

remains unchanged if θ is replaced by $-\theta$. The orbit is therefore symmetrical about the axis passing through O in the direction of the initial line $\theta = 0$. This axis of symmetry is in the direction of the Lenz-Runge vector. Putting $\theta = 0$ we see that the orbit cuts the axis of symmetry at the point P_1 a distance r_1 from O, given by

$$r_1 = \frac{l}{1+e}.$$

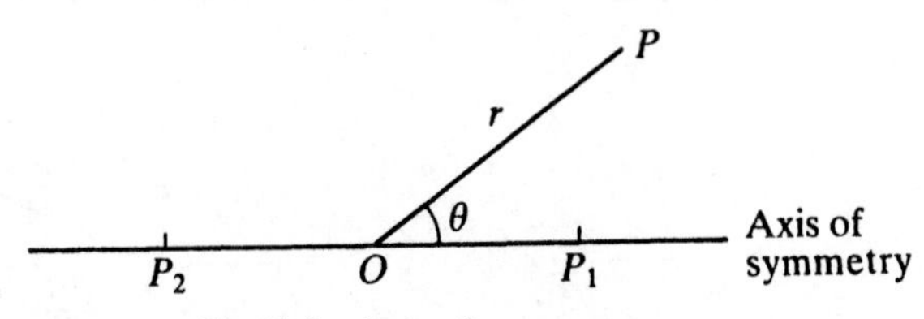

Fig 7.4 Axis of symmetry.

Putting $\theta = \pi$ it would seem that the orbit cuts the axis of symmetry at a second point P_2 a distance r_2 from O, given by

$$r_2 = \frac{l}{1-e}.$$

Care has to be taken here because r_2, being a distance, must be non-negative and so the second point P_2 will exist only when $e < 1$. For $e > 1$ the orbit cuts the axis of symmetry once only and in the limiting case $e = 1$ the orbit can be thought of as cutting the axis twice, the second point P_2 being at infinity. The polar equations of the circular, closed and open orbits identified in Table 7.1 correspond to values of the eccentricity satisfying $e = 0$, $0 < e < 1$ and $e \geq 1$, respectively, the value $e = 1$ giving the limiting example of a closed orbit. We shall discuss these four different cases separately.

Case 1 $e = 0$

With $e = 0$ the polar equation of the orbit yields $r = l$ and describes a circle of radius l. In order to rederive the result obtained in Example 3 of Chapter 6 we use the equations

$$mr^2\dot{\theta} = L \quad \text{and} \quad l = \frac{L^2}{Gm^2m'}.$$

The circular orbit has radius $r = l$ and angular velocity $\dot{\theta} = \omega$. It follows that $L = ml^2\omega$ so that

$$l = \frac{m^2l^4\omega^2}{Gm^2m'}.$$

From this it follows that the radius of the circular orbit is $\sqrt{[3]Gm'/\omega^2}$, as required.

Case 2 $0 < e < 1$

In this case the orbit cuts the axis of symmetry twice, at the points P_1 and P_2. Since

$$r_1 = \frac{l}{1+e} \quad \text{and} \quad r_2 = \frac{l}{1-e}$$

it follows that $r_1 < r_2$ so that these points are those identified in Fig 7.2. In the cases of planetary motion P_1 is the perihelion and P_2 the aphelion. Notice that the fixed particle O is not at the centre C of the line segment P_1P_2. It is instructive to investigate the polar equation of the orbit rewritten in terms of polar coordinates r', θ' defined relative to a new origin O' with O and O' equidistant from the centre C as shown in Fig 7.5. The calculation is rather tedious and is given in Example 1 at

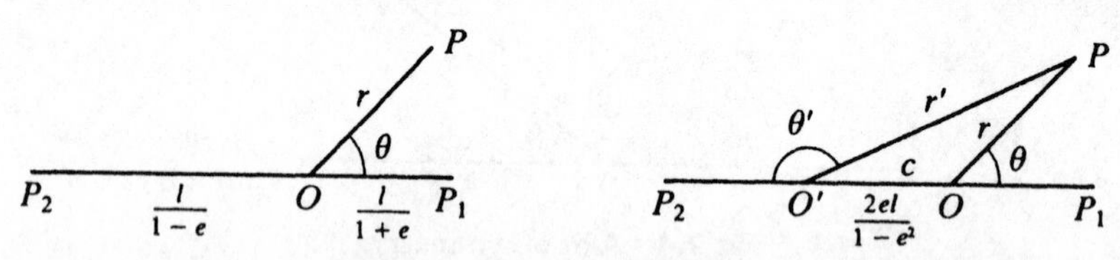

Fig 7.5 The new origin O'.

the end of this Section. The conclusion is somewhat surprising; the polar equation of the orbit is unchanged! This means that there is a second axis of symmetry, perpendicular to the first and passing through the point C. The orbit is illustrated in Fig 7.6. It is in the shape of an ellipse and is called an **elliptic orbit**. The point C is called the **centre** and the points O and O' are called the **foci**. The largest diameter

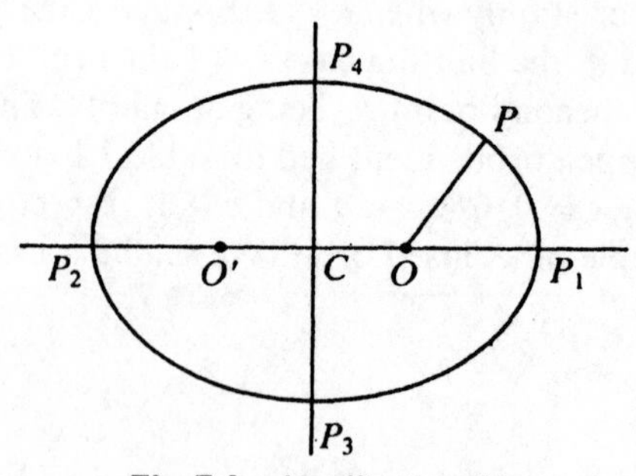

Fig 7.6 An elliptic orbit.

P_1P_2 is called the **major axis** of the elliptic orbit and the smallest diameter P_3P_4 is called the **minor axis**. Half the lengths of these axes, that is the lengths of the semimajor and semiminor axes are conventionally denoted by a and b, respectively. From Fig 7.5

$$a = \frac{1}{2}\left(\frac{l}{1-e} + \frac{l}{1+e}\right) = \frac{l}{1-e^2}.$$

TUTORIAL PROBLEM 7.3

Find the distance of O from C and hence show that the polar coordinates of the point P_3 satisfy

$$r \cos\theta = -\frac{le}{1-e^2}.$$

By substituting into the polar equation of the orbit obtain an expression for r and then, by applying Pythagoras's Theorem to a suitable triangle, prove that the length of the semiminor axis of the elliptic orbit is

$$b = \frac{l}{(1-e^2)^{1/2}}.$$

Case 3 $e > 1$

In this case the orbit cuts the axis of symmetry once only, at the point P_1 with

$$r_1 = \frac{l}{1+e}.$$

As $r \to \infty$ it follows from the polar equation

$$\frac{1}{r} = \frac{1}{l}(1 + e\cos\theta)$$

that $\cos\theta \to -1/e$. The orbit therefore has two asymptotes and is illustrated in Fig 7.7. It is in the shape of one branch of a hyperbola and is called a **hyperbolic orbit**. The point O is called the **focus**. Readers familiar with the hyperbola as a two branched curve will know that there are two foci, as in the case of the ellipse. It is reassuring that the orbit is only one branch of the hyperbola since the two branches are disjoint and the particle could not move on both!

Case 4 $e = 1$.

This is very similar to Case 3, the asymptotes being parallel to the axis of symmetry. The orbit is illustrated in Fig 7.8. It is in the shape of a parabola and is called a **parabolic orbit**. The point O is called the **focus** and P_1 the **vertex**. Many readers will be familiar with the cartesian equation of a parabola. To obtain the cartesian equation of the orbit illustrated in Fig 7.8, take cartesian axes in the plane of motion with origin at the vertex P_1 and y axis directed along the axis of

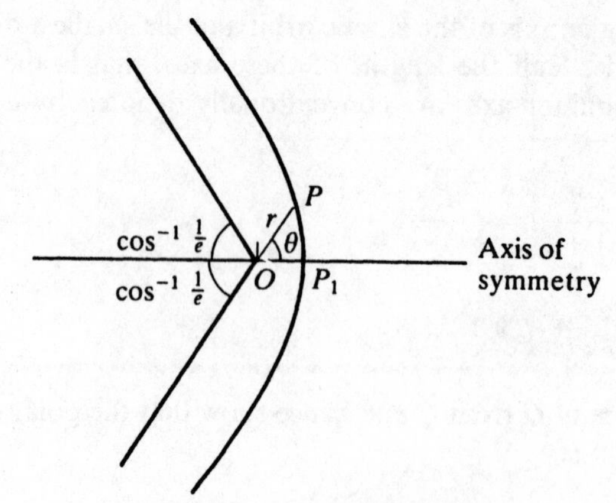

Fig 7.7 A hyperbolic orbit.

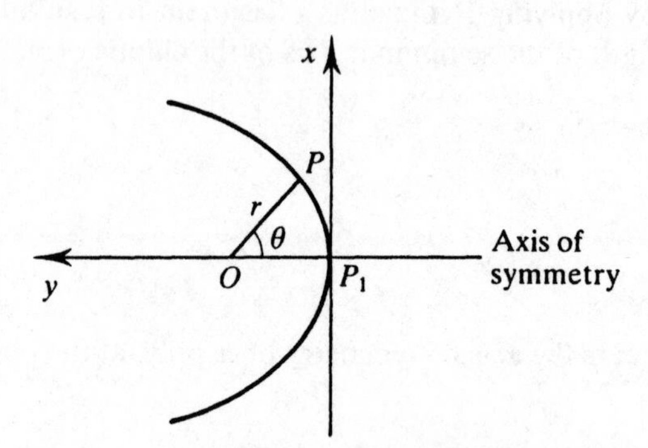

Fig 7.8 A parabolic orbit.

symmetry towards the focus O, as illustrated. Now

$$y = OP_1 - r\cos\theta$$

and $\quad x = r\sin\theta.$

Since $e = 1$, $OP_1 = l/2$ and the polar equation of the orbit becomes

$$\frac{1}{r} = \frac{1}{l}(1 + \cos\theta).$$

It follows that $r\cos\theta = l - r$ so that

$$y = \frac{l}{2} - l + r = r - \frac{l}{2}$$

and $\quad x^2 = r^2\sin^2\theta = r^2 - r^2\cos^2\theta = r^2 - (l - r)^2 = l(2r - l).$

Hence

$$y = \frac{1}{2l}x^2$$

which you may recognize as the cartesian equation of a parabola.

The circle, ellipse, hyperbola and parabola constitute a family of curves called **conics**. Each is obtained by taking a plane section of a complete circular cone; in this context the family of curves are called **conic sections**. You will probably be relieved to learn that problems in orbit theory can readily be solved without any detailed prior knowledge of the geometry of these curves. Such problems often involve the determination of the orbit given some initial conditions. If the initial conditions specify the velocity for a given value of r then they can be used to determine the values of the constants L and E and hence to find the semilatus rectum l and eccentricity e using the expressions

$$l = \frac{L^2}{Gm^2m'} \tag{1}$$

and $$e = \left(\frac{2L^2E}{G^2m^3m'^2} + 1\right)^{1/2} \tag{2}$$

If the initial conditions involve a given value of θ then they must be substituted into the polar equation of the orbit. This equation can only be taken in its standard form

$$\frac{1}{r} = \frac{1}{l}(1 + e\cos\theta) \tag{3}$$

if the axis of symmetry of the orbit is specified, directly or indirectly, by the problem. Otherwise, the equation has to be written as

$$\frac{1}{r} = \frac{1}{l}(1 + e\cos(\theta + \delta)), \tag{4}$$

where δ is the angle between the axis of symmetry and the initial line of the polar coordinate system being used.

The conditions on the eccentricity e for the different types of conic are summarized in Table 7.2. Using the equation (2) above these conditions are also given in terms of the total energy E of the orbiting particle and coincide with the conditions given in Table 7.1 for the various types of orbit.

e	Type of conic	Conditions on E
$e = 0$	circle	$E = -Gm^3m'^2/2L^2$
$0 < e < 1$	ellipse	$0 > E > -Gm^3m'^2/2L^2$
$e = 0$	parabola	$E = 0$
$e > 1$	hyperbola	$E > 0$

Table 7.2 Types of conic

Example I

Using Figure 7.5, show that the polar equation of the orbit of P, namely

$$\frac{1}{r} = \frac{1}{l}(1 + e \cos \theta),$$

becomes

$$\frac{1}{r'} = \frac{1}{l}(1 + e \cos \theta')$$

when written in terms of the polar coordinates r', θ' of P relative to the new origin O'.

SOLUTION

The coordinate r' can be found in terms of r and θ by applying the cosine formula to the triangle OPO'. Thus

$$r'^2 = r^2 + \frac{4e^2l^2}{(1-e^2)^2} + \frac{2r2el}{1-e^2}\cos \theta.$$

From the polar equation of the orbit of P it follows that

$$er \cos \theta = l - r$$

so that

$$\begin{aligned} r'^2 &= r^2 + \frac{4e^2l^2}{(1-e^2)^2} + \frac{4l(l-r)}{1-e^2} \\ &= r^2 + \frac{4l^2}{(1-e^2)^2} - \frac{4lr}{1-e^2} \\ &= \left(r - \frac{2l}{1-e^2}\right)^2. \end{aligned}$$

Hence

$$r' = \pm\left(r - \frac{2l}{1-e^2}\right),$$

where the sign has to be chosen to make $r' \geq 0$. Now

$$\frac{2l}{1-e^2} = \frac{2l}{(1+e)(1-e)}.$$

From $e < 1$ it follows that $1 + e < 2$ and so

$$\frac{2l}{1-e^2} > \frac{l}{1-e}.$$

Now $l/(1-e)$ is the greatest distance of the orbiting particle from O and so

$$\frac{2l}{1-e^2} > r.$$

It follows that the negative sign must be chosen above so that

$$r' = \frac{2l}{1-e^2} - r$$

or, written more symmetrically,

$$r + r' = \frac{2l}{1-e^2}.$$

From Fig 7.5 it also follows that

$$r \cos\theta = -r' \cos\theta' - \frac{2el}{1-e^2}$$

and using these results to eliminate r and $r \cos\theta$ from the equation $er \cos\theta = l - r$ gives

$$-er' \cos\theta' - \frac{2e^2 l}{1-e^2} = l - \frac{2l}{1-e^2} + r'.$$

Simplifying and multiplying through by -1 yields $er' \cos\theta' = l - r'$ so that

$$\frac{1}{r'} = \frac{1}{l}(1 + e\cos\theta').$$

•

The result $r + r' = 2l/(1-e^2)$ obtained in the above Example proves that the sum of the distances from the foci to a point lying on an ellipse is constant. This is a characteristic property of ellipses which can be used to mark out ornamental flower beds, etc.

Example 2

Show that the angular momentum of a planet is related to the perihelion and aphelion distances r_1 and r_2 by the equation $L^2 = 2Gm^2m'r_1r_2/(r_1+r_2)$.

SOLUTION

Eliminating e between the two equations

$$r_1 = \frac{l}{1-e} \quad \text{and} \quad r_2 = \frac{l}{1+e}$$

gives

$$l = \frac{2r_1r_2}{r_1+r_2}$$

so that, using equation (1) above,

$$L^2 = \frac{2Gm^2m'r_1r_2}{r_1+r_2}.$$

Note that the expressions for r_1 and r_2 need not be committed to memory, they are easily obtained by substituting $\theta = 0$ and $\theta = \pi$ into the polar equation of the orbit.

•

Example 3

A comet is observed at a distance of 1.24×10^{11}m from the sun, travelling towards the sun with a speed $5.62 \times 10^4 \text{ms}^{-1}$ at an angle of 45° with its position vector relative to the sun. Obtain the polar equation of the orbit of the comet with the origin at the sun and initial line passing through the observed position of the comet.

SOLUTION

The axis of symmetry is not specified in this problem, so that we must take the polar equation of the orbit in the form

$$\frac{1}{r} = \frac{1}{l}(1 + e\cos(\theta + \delta))$$

and find values for L,e and δ from the initial conditions $\theta = 0$, $r = 1.24 \times 10^{11}$, $\dot{r} = 5.62 \times 10^4 \cos 45°$, $r\dot{\theta} = 5.62 \times 10^4 \sin 45°$. Substituting these into the polar equation gives

$$\frac{1}{1.24 \times 10^{11}} = \frac{1}{l}(1 + e\cos\delta).$$

Differentiating the polar equation of the orbit with respect to t we have

$$-\frac{1}{r^2}\dot{r} = -\frac{e}{l}\sin(\theta + \delta)\dot{\theta}$$

and substituting the initial conditions into this equation gives

$$\frac{5.62 \times 10^4 \cos 45°}{1.24 \times 10^{11}} = \frac{e\sin\delta}{l} 5.62 \times 10^4 \sin 45°.$$

A third equation is required to evaluate the three unknowns. This is obtained by substituting the initial conditions into the equation (1) above with $L = mr^2\dot{\theta}$, to give

$$l = \frac{1.24^2 \times 10^{22} \times 5.62^2 \times 10^8}{Gm'2}.$$

Taking $G = 6.67 \times 10^{-11}\text{m}^3\text{kg}^{-1}\text{s}^{-2}$ and $m' = 1.99 \times 10^{30}$kg gives

$$l = 1.83 \times 10^{11}\text{m}.$$

Substituting this value into the previous two equations yields

$$e\cos\delta = .475 \quad \text{and} \quad e\sin\delta = 1.48.$$

Hence

$$e = 1.55 \quad \text{and} \quad \delta = 72°.$$

●

Example 4

A comet approaches the sun from afar moving with speed v parallel to the axis $\theta = 0$ and at a perpendicular distance d from the axis. Prove that the magnitude of the angular momentum of the comet is given by $L = mdv$ and that the eccentricity is given by $e^2 = 1 + d^2v^4/G^2M^2$, where M is the mass of the sun and m the mass of the comet.

SOLUTION

The angular momentum is the moment of the linear momentum and is therefore the magnitude of the linear momentum multiplied by the perpendicular distance of the origin from the path of the comet. As the comet approaches from afar its linear momentum has magnitude mv and the perpendicular distance is d, so that $L = mdv$. The energy equation with $|\dot{\mathbf{r}}| = v$ and $r \to \infty$ gives

$$E = \frac{1}{2}mv^2$$

so that, using the expression (2) above for e in terms of E, gives

$$e^2 = 1 + \frac{L^2mv^2}{G^2m^3M^2}$$

or, eliminating L,

$$e^2 = 1 + \frac{d^2v^4}{G^2M^2}.$$

●

Example 5

At the end of a launching process a satellite is travelling with velocity v parallel to the earth's surface and at a height h. Prove that if $(R+h)v^2 > Gm'$ the eccentricity of its orbit is given by

$$1 + e = \frac{(h+R)v^2}{Gm'},$$

where m' and R are the mass and radius of the earth. Show that the greatest height H of the satellite is given by

$$H = \frac{(R+h)^2v^2}{2Gm' - (R+h)v^2} - R.$$

SOLUTION

By hypothesis, at the end of the launching process the satellite has no radial velocity and therefore is located at either the perigee or apogee. We can take the polar equation of its orbit in the standard form (3) above but cannot determine whether $\theta = 0$ or π at the end of the launching process. The initial conditions are therefore

$$\theta = 0 \quad \text{or} \quad \pi, \qquad r = R + h, \qquad \dot{r} = 0, \qquad r\dot{\theta} = v.$$

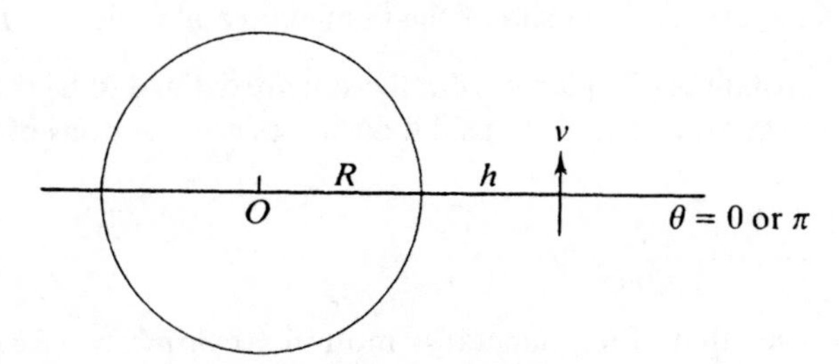

Fig 7.9 Launch of satellite.

Using these $L = mr^2\dot{\theta} = m(R+h)v$ so that, using equation (1) above, $l = L^2/Gm^2m' = (R+h)^2v^2/Gm'$. Substituting into the standard polar equation of the orbit gives

$$\frac{1}{R+h} = \frac{.G'm}{(R+h)^2v^2}(1 \pm e)$$

or $$1 \pm e = \frac{(R+h)v^2}{Gm'}.$$

We are given that $(R+h)v^2 > Gm'$ and, remembering that $e \geq 0$, it follows that the positive sign must be taken. The end of the launching process is therefore at the perigee and the required maximum height H of the satellite will be at the apogee when $\theta = \pi$. Hence

$$\frac{1}{R+H} = \frac{Gm'}{(R+h)^2v^2}(1-e) = \frac{Gm'}{(R+h)^2v^2}\left[2 - \frac{(R+h)v^2}{GM'}\right]$$

which leads to the required result. •

EXERCISES ON 7.2

The following data should be used, if required, when answering these and subsequent exercises. Value of the universal gravitational constant $G: 6.67 \times 10^{-11}\text{m}^3\text{kg}^{-1}\text{s}^{-2}$; mass and radius of the sun: 1.99×10^{30}kg and 6.96×10^8m respectively; mass and radius of the earth: 5.98×10^{24}kg and 6.37×10^6m respectively.

1. Draw the graph of the effective potential energy $U(r)$ of a particle of mass m moving with angular momentum $\mathbf{L}$ under the action of an attractive central force $F(r) = -\lambda/r^4$. Deduce that the particle can move in a circle of radius $\lambda m/L^2$. Discuss the stability of this circular orbit and compare the situation with that of motion under the action of an attractive inverse square force. Generalize your results by proving that the circular orbits of a particle moving under an attractive central force $F(r) = -\lambda/r^n$ are stable or unstable depending on whether $n < 3$ or $n > 3$.

2. Show that the eccentricity of an ellipse can be written as

$$e = \frac{r_2 - r_1}{r_2 + r_1},$$

where r_1 and r_2 are the distances of the perihelion and aphelion from the sun.

3. The apsidal distances of a planet from the sun are defined to be those values of r for which $\dot{r} = 0$. Show that the apsidal distances are solutions of the quadratic equation

$$Er^2 + Gmm'r - \frac{L}{2m} = 0.$$

Hence deduce that for planetary motion $G^2m^3m'^2 > -2EL^2$ and that $E = -Gmm'/(r_1 + r_2)$, where r_1 and r_2 are the two solutions of the equation.

4. A comet is observed at a distance 1.32×10^{11}m from the sun, travelling away from the sun with a speed 4.5×10^4ms^{-1} at an angle of 30° to the radius from the sun. Will the comet ever again approach the sun?

5. A comet approaches the sun, its distance from the sun at the perihelion being observed to be r_1. When the position vector to the comet has turned through an angle of 90° from the perihelion, the distance of the comet from the sun is observed to be r_0. Show that the comet will escape from the solar system (i.e. $e \geq 1$) if and only if $r_0 \geq 2r_1$.

6. A satellite moves in a circular orbit with speed v. A rocket gives a boost to the satellite increasing its speed by an amount δv in the direction of motion of the satellite. Show that, for small δv, the eccentricity e of the satellite's new orbit is given by $e = 2\delta v/v$. Where does the apogee of the new orbit now occur? The reverse process, i.e. firing a retro-rocket to change an elliptic orbit into a circular orbit, is used before landing space shuttles.

7.3 Kepler's Laws of Planetary Motion

In 1609 Kepler published his three laws of planetary motion, found by analysing the meticulous observations made by the astronomer Tycho Brahe before the advent of the telescope. These laws describe the kinematics of planetary motion and from them Newton was able to deduce his model of gravitation and hence investigate the mechanics of planetary motion. The three laws can be stated as:

- Kepler's First Law: each planet moves on an ellipse with the sun at one focus
- Kepler's Second Law: areas swept out by the line segment joining the sun to the planet in equal times are equal
- Kepler's Third Law: the square of the period of revolution of the planet, that is the planetary year, is proportional to the cube of the length of the semi major axis, the constant of proportionality being the same for all planets.

These three laws can be deduced from the theory of orbits discussed in Section 7.2, it being assumed that each planet orbits about the sun which is fixed in some inertial frame.

The first law is deduced trivially because the planets move on closed orbits and the only possible closed orbits are elliptical with the fixed particle, that is the sun, at one focus.

Since the angular momentum of the orbiting particle, that is the planet, is conserved it follows from Example 6 of Chapter 4 that the rate of change of the areas swept out by the given line segment is constant, equal in value to $|\mathbf{L}|/2m$. This is an alternative and simpler statement of the second law; of course Kepler himself could not have stated the law in terms of a rate of change because the calculus had not been developed in the early seventeenth century.

Integrating the equation

$$\frac{dA}{dt} = \frac{|\mathbf{L}|}{2m},$$

obtained in Example 6 of Chapter 4, with respect to time over a complete orbit of the planet yields

$$A = \frac{|\mathbf{L}|}{2m} T,$$

where A is the area enclosed by the orbit and T is the time taken for a complete orbit, that is the period of revolution or planetary year. The orbit is an ellipse and the area enclosed by an ellipse with semimajor and minor axes of lengths a and b respectively is πab. Hence

$$\pi ab = \frac{|\mathbf{L}|}{2m} T$$

so that

$$T^2 = \frac{4m^2\pi^2 a^2 b^2}{|\mathbf{L}|^2}.$$

In Tutorial Problem 7.3 we saw that $b = l/(1 - e^2)^{1/2}$. Hence

$$b^2 = \frac{l^2}{1 - e^2} = al = \frac{aL^2}{Gm^2 m'},$$

where we have used equation (1) above. Now $L^2 = |\mathbf{L}|^2$ and it follows that

$$T^2 = \frac{4\pi^2}{Gm'} a^3.$$

This is Kepler's third law, the constant of proportionality, $4\pi^2/Gm'$, being independent of the planet.

The elliptic orbits of the planets found above are in fact only approximations to the observed orbits. This is only to be expected because several assumptions have been made in the modelling process. First it was assumed that the sun is fixed in an inertial frame. This is corrected in the next Section. Secondly the effect of the gravitational attraction of other planets on the orbiting planet has been neglected. Since the masses of the planets are very small compared to the mass of the sun they produce only small deviations to the elliptic orbits. These small deviations are important however. For example in the early eighteen hundreds the observed orbit of Uranus showed a deviation from the orbit calculated using Newton's equation of motion even after the gravitational attractions of all the other known planets had been taken into account. Adams and Leverrier found that this deviation could be explained by the presence of an unknown planet and the planet which we now know as Neptune was indeed found in the predicted position! A similar deviation is found in the case of Mercury. A slow rotation, or precession, of the major axis of the elliptic orbit of Mercury remains after taking into account the gravitational attraction of all other known planets. This residual rotation amounts to 41 seconds of arc per century. The cause of this rotation was a mystery until 1917 when Albert Einstein published his theory of general relativity. This theory predicts just such a rotation. More recently it has been suggested that part of this residual rotation is due to the oblateness of the sun, a conjecture which is hard to verify simply because the brightness of the sun makes it hard to observe its precise shape. One mathematical feature of planetary orbits about an oblate sun is worth mentioning; the force of attraction is only a central force for orbits lying in the equatorial plane of the sun!

Example 6

On 12th April, 1961 Y.A.Gagarin made the first manned space flight. The perigee and apogee of his craft (Vostik 1) were 1.81×10^5m and 3.27×10^5m above the earth's surface. Calculate the period of revolution of the space craft.

SOLUTION

The distances r_1 and r_2 of perigee and apogee from the centre of the earth are

$$r_1 = 3.27 \times 10^5 + 6.37 \times 10^6 = 6.70 \times 10^6\text{m}$$

and

$$r_2 = 1.81 \times 10^5 + 6.37 \times 10^6 = 6.55 \times 10^6\text{m}.$$

The sum of these distances is twice the length of the semimajor axis so that $a = 6.63 \times 10^6$m. Substituting this into Kepler's third law gives

$$T^2 = \frac{4\pi^2 \times 6.63^3 \times 10^{18}}{6.67 \times 10^{-11} \times 5.98 \times 10^{24}}.$$

From this $T = 89$ minutes. •

TUTORIAL PROBLEM 7.4

Gagarin's flight lasted 108 minutes. What factors might explain the discrepancy between this and the calculated period of revolution of the spacecraft?

EXERCISES ON 7.3

1. Show that the mass of a planet can be determined if the period of revolution and the semi major axis of one of its satellites are known. Ganymede, a satellite of the planet Jupiter, has a period of 7.155 days and moves on an almost circular orbit of radius 1.071×10^9m. Find an estimate for the mass of Jupiter.

2. Halley's comet was last seen in 1986 moving on a very eccentric orbit with $e = 0.9674$, its perihelion being a distance of 8.77×10^{10}m from the sun. Halley was born in 1656; how old was he when he discovered the comet? When will the comet next visit the solar system?

7.4 The Two Body Problem

In this section we will discuss the motion of two particles P and P' of masses m and m' moving under the action of their mutual gravitational attractions alone.

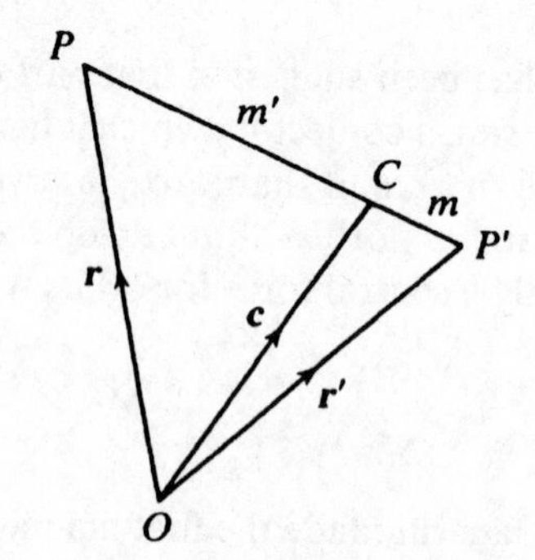

Fig 7.10 Two orbiting particles.

Following the discussion at the beginning of Section 7.2 this particle motion can be used, in appropriate circumstances, to model the motion of two bodies of finite size or, more generally, the motion of the centres of two spherically symmetric bodies, for example the two components of a binary star. Let $\mathbf{r}$ and $\mathbf{r}'$ be the position vectors of the particles relative to an origin O fixed in an inertial frame S. The equations of motion of the two particles are

$$m\ddot{\mathbf{r}} = Gmm'\frac{(\mathbf{r}' - \mathbf{r})}{|\mathbf{r}' - \mathbf{r}|^3}$$

and $$m'\ddot{\mathbf{r}}' = Gmm'\frac{(\mathbf{r} - \mathbf{r}')}{|\mathbf{r}' - \mathbf{r}|^3}.$$

Adding gives

$$m\ddot{\mathbf{r}} + m'\ddot{\mathbf{r}}' = 0.$$

The **centre of mass** of the two particles is defined to be the point C with position vector $\mathbf{c}$ given by

$$\mathbf{c} = \frac{m\mathbf{r} + m'\mathbf{r}'}{m + m'}.$$

It follows from the previous equation that $\ddot{\mathbf{c}} = 0$, so that the centre of mass C moves on a straight line with constant speed relative to the inertial frame S. A second inertial frame can therefore be chosen with origin at the centre of mass C. We will denote this frame by $\bar{S}$.

Using the ratio theorem of vector algebra we can interpret the centre of mass C as being the point lying on the line segment PP' and dividing it in the ratio $m' : m$. If the mass of one of the particles, say P', is very much greater than the mass of the other then the centre of mass will be very close to the position of P'. In these circumstances P' can be modelled as actually being located at the centre of mass and so fixed in the inertial frame $\bar{S}$. This justifies the model used in the previous Sections. A numerical example is quite instructive.

Example 7

Show that the centre of mass of the sun and earth lies within the sun's surface.

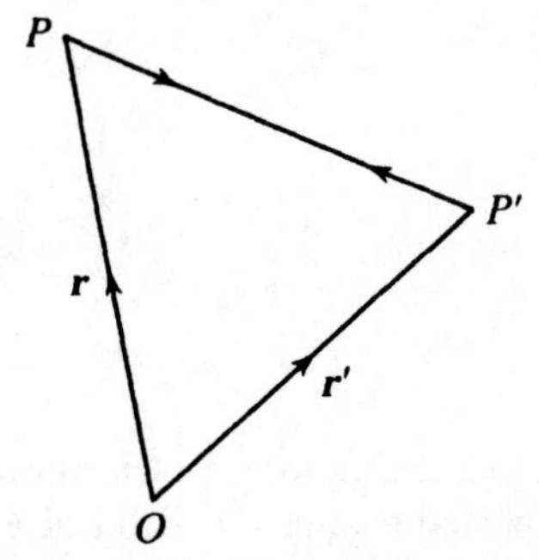

Fig 7.11 The centre of mass.

SOLUTION

The average distance between the earth and the sun is approximately 15×10^{10}m, all other relevant data is given at the beginning of Exercises 7.2. Let x_c be the distance of the centre of mass from the centre of the sun. Then substituting into the one dimensional form

$$x_c = \frac{mx + m'x'}{m + m'}$$

of the equation defining the centre of mass, with the centre of the sun as origin, gives

$$\begin{aligned} x_c &= \frac{5.98 \times 10^{24} \times 15 \times 10^{10}}{1.99 \times 10^{30} + 5.98 \times 10^{24}} \\ &= 4.5 \times 10^4\text{m}. \end{aligned}$$

The radius of the sun is 6.96×10^8m and so the centre of mass does indeed lie within the surface of the sun. •

If the mass of neither particle is very much greater than the mass of the other then it is no longer possible to think of one particle as being fixed with the other particle orbiting about it. Instead, each particle can be considered as orbiting about their fixed centre of mass C. Relative to C the equation of motion of P is

$$m\ddot{\bar{\mathbf{r}}} = Gmm'\frac{(\bar{\mathbf{r}}' - \bar{\mathbf{r}})}{|\bar{\mathbf{r}}' - \bar{\mathbf{r}}|^3},$$

where the bar $\bar{}$ is used to denote the position vectors relative to C. Of course $\bar{\mathbf{c}} = 0$ so that

$$m\bar{\mathbf{r}} + m'\bar{\mathbf{r}}' = 0.$$

Using this equation to eliminate $\bar{\mathbf{r}}'$ from the equation of motion of P yields

$$m\ddot{\bar{\mathbf{r}}} = -Gmm'\left(\frac{m'}{m + m'}\right)^2 \frac{\bar{\mathbf{r}}}{|\bar{\mathbf{r}}|^3}.$$

Putting

$$\bar{\mathbf{r}} = \left(\frac{m'}{m+m'}\right)^{2/3} \mathbf{r}$$

this equation can be rewritten as

$$m\ddot{\bar{\mathbf{r}}} = -Gmm' \frac{\bar{\mathbf{r}}}{|\bar{\mathbf{r}}|^3}.$$

This last equation is identical to the equation of motion of P which would have held had P' been fixed in an inertial frame. The orbit of P relative to C is therefore similar to the orbits discussed in the last two sections, the only difference being that the size of the orbit is reduced by the scaling factor

$$\left(\frac{m'}{m+m'}\right)^{2/3}.$$

In calculating the orbit of the earth this scaling factor is equal to one, within the accuracy of the data given at the beginning of Exercises 7.2. This again indicates the accuracy of modelling the sun as being fixed in an inertial frame.

The discussion of the orbit of P relative to C carries over to the orbit of P' relative to C, the only difference being that the scaling factor is ...... perhaps you can guess? An alternative approach to this two body problem is to still think of P as orbiting about P', but with P' no longer fixed. To do this consider the position vector $\mathbf{R} = \mathbf{r} - \mathbf{r}'$ of P relative to P'. Using the two equations of motion

$$\ddot{\mathbf{R}} = \ddot{\mathbf{r}} - \ddot{\mathbf{r}}' = Gm' \frac{(\mathbf{r}' - \mathbf{r})}{|\mathbf{r}' - \mathbf{r}|^3} - Gm \frac{(\mathbf{r} - \mathbf{r}')}{|\mathbf{r}' - \mathbf{r}|^3}$$

$$= -G(m+m') \frac{|\mathbf{R}|}{|\mathbf{R}|^3}.$$

This is usually written in the form

$$\frac{mm'}{m+m'} \ddot{\mathbf{R}} = -Gmm' \frac{\mathbf{R}}{|\mathbf{R}|^3}$$

and is again similar to the equation of motion of P which would hold had P' been fixed in an inertial frame, the only difference being that the inertial mass of P is replaced by $mm'/(m+m')$. This is known as the **reduced mass**.

EXERCISES ON 7.4

1. The moon moves in an almost circular orbit of radius 3.84×10^8m and is of mass 7.35×10^{22}kg. Find the distance of the centre of mass of the earth and moon from the centre of the earth.

2. Show that the equation of motion of a particle P orbiting about a particle P' can be written in a form analogous to that which would hold if P' were fixed in an inertial frame, the only difference being that the gravitational mass of P' is replaced by $m + m'$.

3. The equation of motion of two particles P_1 and P_2 of masses m_1 and m_2 are

$$m_1\ddot{\mathbf{r}}_1 = \mathbf{F}_1 + \mathbf{F}_{12}$$

$$\text{and} \qquad m_2\ddot{\mathbf{r}}_2 = \mathbf{F}_2 + \mathbf{F}_{21},$$

where $\mathbf{r}_1$ and $\mathbf{r}_2$ are the position vectors of the particles relative to an inertial frame, $\mathbf{F}_{12}$ and $\mathbf{F}_{21}$ are forces acting between the particle satisfying Newton's third law and $\mathbf{F}_1$ and $\mathbf{F}_2$ are other external forces acting on the particles. Prove that the centre of mass of the two particles moves as a particle of mass $m_1 + m_2$ acted on by a force $\mathbf{F}_1 + \mathbf{F}_2$.

Summary

- the force $\mathbf{F} = F(r)\hat{\mathbf{r}}$ is a **central force** and is **attractive** if $F(r) < 0$ and **repulsive** if $F(r) > 0$
- the angular momentum $\mathbf{L}$ of a particle moving under a central force is conserved so that the path of the particle lies on a plane
- for an **inverse square force,** $F(r) = -\lambda/r^2$
- for inverse square forces alone a second conserved vector exists, called the Lenz-Runge vector and given by $\mathbf{k} = \dot{\mathbf{r}} \times \mathbf{L} - \lambda\mathbf{r}/r$.
- the different types of orbit for a particle P orbiting about a particle P' which is fixed in an inertial frame can be investigated using the **effective potential**

$$U(r) = \frac{L^2}{2mr^2} - \frac{Gmm'}{r}.$$

- the polar equation of the orbit of the particle P is

$$\frac{1}{r} = \frac{1}{l}(1 + e\cos\theta)$$

where l is the **semilatus rectum** and e the **eccentricity**
- the type of orbit is determined by the value of e
- for elliptic orbits the prefixes **peri** and **ap** are used to describe the positions of closest approach and greatest retreat from the focus
- the expressions

$$l = \frac{L^2}{Gm^2m'} \qquad \text{and} \qquad e = \left(\frac{2L^2E}{G^2m^3m'^2} + 1\right)^{1/2}$$

are often useful in solving problems
- the kinematics of planetary motion is described by **Kepler's three laws**
- the **centre of mass** of two particles of masses m and m' divides the line segment joining them in the ratio $m' : m$
- if two particles are moving under their mutual gravitational attractions alone then their centre of mass is fixed in an inertial frame

- the orbit of each particle relative to the centre of mass is similar to the corresponding orbit found under the assumption that the other particle is fixed in an inertial frame
- the **reduced mass** is $m_1m_2/(m_1+m_2)$ and replaces the inertial mass of a particle which is orbiting about a second particle which is not fixed in an inertial frame

FURTHER EXERCISES

1. Discuss qualitatively, using the appropriate effective potential, the types of orbits which can occur for motion under an inverse cube central force $F(r) = -k/r^3$, where $k > 0$.

1. Prove that the total energy of a particle moving on an elliptic orbit under the action of an inverse square attraction can be written as $E = -Gmm'/2a$ and hence deduce that the speed v of the particle is given by

$$v^2 = Gm'\left(\frac{2}{r} - \frac{1}{a}\right).$$

3. Show that the total energy E of a planet is related to its kinetic energies T_p and T_a at perihelion and aphelion by the equation

$$E^2 = T_pT_a.$$

4. When $\theta = \theta_0$ the direction of motion of a comet is observed to make an angle of θ_1, in the same sense, with the position vector of the comet relative to the sun. Here θ is measured from the major axis of the orbit of the comet. Prove that the eccentricity is given by

$$e = -\frac{\cos\theta_1}{\cos(\theta_0 + \theta_1)}.$$

5. Perihelion and aphelion of the planet Pluto occur at distances of 4.425×10^{12}m and 7.375×10^{12}m from the sun, respectively. Find the eccentricity of Pluto's orbit and the length of Pluto's planetary year.

8 • Looking Ahead

8.1 The Tip of an Iceberg

Out attention in this text has been confined almost exclusively to the motion of a single body, modelled as a particle and based on the classical mechanics of Newton. This is, indeed, the tip of an iceberg; most of classical mechanics remains for you to study. Section 7.4 introduced you to the two body problem and for the first time we discussed the motion of more than one body, although still with the assumption that both bodies can be modelled as particles. This leads naturally to the study of the motion of a **general system of particles**. If the internal forces acting between the particles of a given system satisfy the weak form of Newton's third law of motion we find that the rate of change of the total linear momentum of the system is equal to the total force acting, that is the sum of the external forces acting on each individual particle of the system. If the internal forces also satisfy the strong form of the third law then we find the rate of change of the total angular momentum of the system about a suitably chosen origin is equal to the total moment acting on each individual particle of the system. These two equations of motion are important because they are independent of any internal forces acting between the particles, the details of which are often unknown. The concept of the **centre of mass** of a system of particles is of great importance and a consequence of the first of the equations of motion is that the centre of mass of the system moves as a particle of mass equal to the total mass of the system moving under the action of a force equal to the total force acting on the system.

If the total force acting on a system of particles is zero then the total linear momentum of the system is conserved. This conservation law is used to discuss the **scattering** of a system of particles, examples range from the collision of billiard balls to the scattering of elementary particles as traced out in a bubble chamber.

When modelling the motion of a **rigid body** of finite size the first question we ask is how can we specify the position of the body as it moves in space? One simple answer would be to specify the coordinates of three non collinear points of the body. We would seem to require nine coordinates in all to specify the position of the body itself. However, the distance between any two points of a rigid body remains constant as the body moves and therefore three equations can be written down relating the three sets of coordinates. We therefore require only six independent coordinates to specify the position of the rigid body in space; the body is said to have **six degrees of freedom**. By dividing a given rigid body into small segments and treating each segment as if it were a particle the body can be modelled as a system of particles. The two vector equations of motion introduced in the first paragraph are sufficient to determine the six independent coordinates, given suitable initial conditions. In writing down, for example, the total linear momentum of the rigid body we have to add together the linear momenta of the small segments into which the body is divided. To ensure that the segments behave as particles we also take the limit as each segment tends to a point. This procedure leads to an integral over the volume occupied by the body. In practice the six independent coordinates introduced above are not the most convenient variables to use and it is often more convenient to use the three coordinates of the centre of mass of the body together with three **eulerian angles** which specify the rotation of the rigid body about its centre of mass; perhaps you can think of how these angles are chosen? The coordinates of the centre of mass are determined by considering the motion of the centre of mass as a particle and the three angles are determined using the equation of motion involving the rate of change of the angular momentum.

We see from the above that the motion of a rigid body can be determined, given suitable initial conditions, provided that the total external force and total moment of the external forces are known. Two different systems of external forces will therefore impart the same motion to a given rigid body if the total forces and the total moments of the two systems are identical. Two such systems of forces are said to be **mechanically equivalent** and this concept is central to the discussion of the **statics** of rigid bodies and more general mechanical systems.

For a general mechanical system it is important to distinguish between the **external forces** acting on the system and the so called **force of constraint** which any connectors, hinges, joints, etc exert on the components which make up the mechanical system. The position of such a system in space is specified by a minimal number of coordinates, called **generalized coordinates**, the number of generalized coordinates being the **number of degrees of freedom** of the system. **Analytical mechanics** is concerned with the determination and discussion of a set of equations of motion which do not involve the forces of constraint but which are sufficient to determine the generalized coordinates and therefore the motion of the mechanical system, given suitable initial conditions. These equations are known as **Lagrange's equations**. This more formal and analytical approach to motion has led to the theory of **dynamical systems**.

The techniques used to discuss the motion of a rigid body are also used in **fluid mechanics** to discuss the motion of liquids and gases. When dividing a fluid into small segments a very important force to consider is the force which neighbouring segments exert on each other; this leads to the concept of **pressure**. The motion of the universe as a whole can be investigated by modelling the galactic clusters as fluid particles. Applying the equations of fluid dynamics then leads to the theory of **Newtonian cosmology** which predicts the possibility, amongst others, of an expanding universe starting from a big bang creation.

The last few paragraphs have been intended to give you just a glimpse of the diversity of the avenues which remain to be studied and we hope that this text will have given you the appetite to continue your study of a subject which can be enjoyed for its own sake and which dominates the applications of mathematics to industry and technology.

8.2 The Newtonian Theory of Ghost Images

Do you think that two bullets from an automatic weapon could hit a given target simultaneously? You might think the question rather silly because surely each bullet will take the same time to travel the distance between the weapon and the target, so that two bullets hitting the target simultaneously would have had to have been fired simultaneously, which is not possible for a single barrelled weapon. However, this response presupposes that the distance between the weapon and the target is fixed, so that the weapon is at rest relative to the target. The situation is far more complex if the weapon is moving relative to the target.

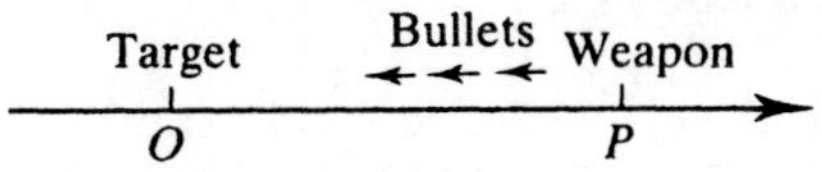

Fig 8.1 A moving weapon.

To investigate the question further we will consider the case of a weapon P moving away from a fixed target O, along a straight line passing through O. Let x be the displacement of P relative to O, the line of motion being orientated as shown in Fig 8.1. Since the weapon is moving relative to the target, x will be a function of time and here we will consider the simple case

$$x = \frac{1}{2}at^2 + ut,$$

where a and u are constants with u positive. This implies that the weapon is moving with a constant acceleration a relative to the target and that at time t the velocity of the weapon relative to the target is given by

$$v = at + u.$$

Suppose that the muzzle speed of the weapon, as specified by the manufacturer, is c. Then the velocity of the bullets relative to the weapon is $-c$, the minus sign appearing because the bullets are moving in the negative x direction. The velocity of that bullet fired at time $t = \tau$, relative to the target, is obtained by applying the usual Newtonian addition law of relative velocities to give

$$\begin{aligned}\text{vel of bullet rel to } O &= \text{vel of bullet rel to } P + \text{vel of } P \text{ rel to } O \\ &= -c + a\tau + u.\end{aligned}$$

Because the bullets are moving in the negative x direction this velocity must be negative. It follows that the bullets approach the target with a speed $c - a\tau - u$. The weapon is located a distance $\frac{1}{2}a\tau^2 + u\tau$ from the target when the bullet is fired and, since the speed is constant, the bullet will take a time

$$\frac{\text{distance}}{\text{speed}} = \frac{\frac{1}{2}a\tau^2 + u\tau}{c - a\tau - u}$$

to reach the target. The bullet fired at the time τ will therefore hit the target at the time T given by

$$T = \tau + \frac{\frac{1}{2}a\tau^2 + u\tau}{c - a\tau - u}.$$

Knowing the time T at which a certain bullet hits the target we can deduce, predict would be the wrong word here, the time τ at which the bullet was fired by solving the above equation for τ. Multiplying both sides by $c - a\tau - u$ and rearranging terms leads to the quadratic equation

$$\frac{1}{2}a\tau^2 - \tau(aT + c) + T(c - u) = 0.$$

The solutions to this equation are

$$\tau = \frac{aT + c \pm \sqrt{(aT + c)^2 - 2aT(c - u)}}{a}.$$

Mathematical models sometimes lead to spurious predictions and this might be the case here because we know that the bullet cannot hit the target before it is fired! Hence the sign in the above must be chosen in order to make $T > \tau$. If $a > 0$ this inequality can be rewritten as

$$aT > a\tau$$

or

$$aT > aT + c \pm \sqrt{(aT + c)^2 - 2aT(c - u)},$$

that is

$$0 > c \pm \sqrt{(aT + c)^2 - 2aT(c - u)}.$$

This can only be satisfied if the negative square root is chosen. It follows that only one time of firing exists for given T and therefore only one bullet can hit the target at time T. If $a < 0$ the inequality $T > \tau$ can be rewritten as

$$aT < a\tau$$

so that

$$0 < c \pm \sqrt{(aT + c)^2 - 2aT(c - u)}.$$

This inequality is satisfied if the positive square root is chosen but will also be satisfied if the negative square root is chosen, provided that

$$\sqrt{(aT + c)^2 - 2aT(c - u)} < c.$$

Squaring and rearranging gives

$$a^2T^2 + 2aTu < 0$$

so that

$$T < -\frac{2u}{a}, \quad \text{i.e. } T < \frac{2u}{|a|}.$$

It follows that if $a < 0$, so that the speed of the weapon is decreasing, and if $T < 2u/|a|$ then there exists two times $\tau = \tau_1$ and $\tau = \tau_2$ of firing for each time T. Hence the two bullets fired at times τ_1 and τ_2 hit the target simultaneously at time T. We can now answer the question asked at the beginning of this section. Yes – two bullets fired from an automatic weapon can hit a given target simultaneously!

You may be wondering what the above has to do with the title of this Section? To answer your query we will consider exactly the same calculation but with the target replaced by an observer O and the weapon by a source of light P. Now photons are "fired" from the source to the observer and the speed c in the calculation becomes the speed of light. If the speed of the source is decreasing so that $a < 0$ then two photons will be received by the observer simultaneously at time T, provided that $T < 2u/|a|$. If these photons are emitted by the source as it passes the points P_1 and P_2 as illustrated in Fig 8.2, then the observer will actually see the source to be at these two points simultaneously! This then is the Newtonian theory of ghost images, the possibility of seeing more than one image of a single object simultaneously.

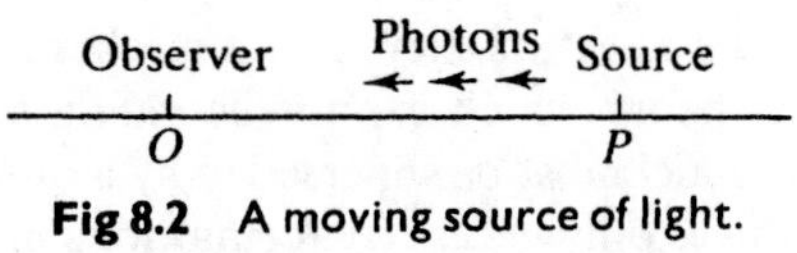

Fig 8.2 A moving source of light.

Ghost images have a property familiar from the ghosts of our childhood, at time $T = 2u/|a|$ the ghost image suddenly disappears! These ghost images have another familiar property, they can pass through closed doors. If the door is at rest relative to the observer then the time taken for a photon to travel from the door will be different to either time taken for a photon to travel from the source to the observer. The door has to be opened in order to allow the physical source of light to pass through it but the observer will not see it open at the same time as the source passes through it. The observer will actually see the source pass through a closed door!

In the early nineteen hundreds the astronomer Willem de Sitter predicted the existence of ghost images of the components of a binary star. Despite painstaking observations made over a considerable period of time he failed to detect such ghost images and therefore to validate this particular prediction of the classical Newtonian theory. What has gone wrong? The mathematics used in this section has been straightforward enough and surely no mistake has been made? What results from the Newtonian theory have been used? If you look at the calculations carefully you will find that the only result we have used is the Newtonian addition law of relative velocities. Is it possible that this familiar law is in fact false? Referring back to Section 2.1 the addition law of relative velocities is based on the addition law of relative displacements and the universality of time. Since ghost images are not observed one or both of these basic assumptions of Newtonian mechanics must be false. According to Einstein's Theory of Special Relativity the universality of time does indeed break down. You might have read that given two identical clocks, if one moves relative to the other then it runs slow so that identical clocks, once synchronized, will not necessarily remain synchronized.

Suppose that, for motion on a straight line, v and v' are the velocities of a particle P relative to two origins O and O' and V is the velocity of O relative to O'. Then the Newtonian addition law of relative velocities

$$v = v' + V$$

is replaced in Special Relativity by the addition law

$$v = \frac{v' + V}{1 + v'V/c^2}.$$

Notice that if v' is equal to the speed of light c then

$$v = \frac{c + V}{1 + cV/c^2} = c$$

so that the speed of light is independent of the choice of the observer. The photons in the above calculation will therefore travel with the same speed relative to both source and observer. There is then no question of the existence of ghost images; in fact de Sitter's failure to observe ghost images is just one validation of the relativistic mechanics based on Special Relativity.

Perhaps it is unusual to end a text on Newtonian mechanics with an example of a prediction based on the theory which cannot be validated and which therefore indicates that Newton's model must be superseded by a more sophisticated one. It should be stressed that this is only necessary for motions in which the speed of the particle approaches the speed of light. Everyday motions can still be modelled using Newtonian theory and as indicated in the last section a great deal of classical mechanics remains for you to study. The purpose of this Section has been to indicate that this study can also be extended to other fascinating and diverse theories of mathematical physics.

Appendix – Learning Resources

Books about students conceptions of mechanics:

Orton, A., *Studies in Mechanics Learning*. Centre for Studies in Science and Mathematics Education, University of Leeds, 1989.

Warren, J.W., *Understanding Force*. John Murray, 1979 (revised edition 1984).

Books using mechanics to model real situations:

Alexander, R.M., *Exploring Biomechanics: Animals in Motion*. Scientific American Library, 1992.

Alexander, R.M., *The Human Mechanic*. Natural History Museum Publications, London, 1992.

Daish, C.B., *The Physics of Ball Games*. Hodder and Stoughton, 1981.

Dyson, G., *Mechanics of Athletics*. Hodder and Stoughton, 1986 (8th edition, revised by Woods, B.D. and Travers, P.R.).

Savage, M. and Williams, J., *Mechanics in Action: Modelling and Practical Investigations*. Cambridge University Press, 1990.

Books for further study:

Berger, V. and Olson, M., *Classical Mechanics – a Modern Perspective*. McGraw Hill, 1973.

French, A.P., *Newtonian Mechanics*. W.W. Norton, 1971.

Symon, K.R., *Mechanics*. Addison Wesley, 1971.

Apparatus:

The Leeds Mechanics Kit. Unilab of Blackburn (includes handbook).

The Oxford Mechanics Kit. Technical Prototypes: 2, New Park Street, Leicester (includes handbook).

Videos:

Mechanics in Action

Modelling Circular Motion

Modelling Satellite Motion

available from the Mechanics in Action Project, School of Mathematics, The University of Leeds.

Computer Software and Video Discs.

The availability of computer software changes rapidly and for up to date information you are advised to contact

Computers in Teaching Initiative
Centre for Mathematics and Statistics
Centre for Computer-Based Learning
University of Birmingham
Birmingham B15 2TT.

The Centre Manager, Pam Bishop, will be happy to answer queries on 0121-414-4800.

Two examples of currently available material are

Motion: A Visual Database

– an interactive video disc available from Media Productions, Anglian Polytechnic University, East Road, Cambridge CB1 1P7.

Interactive Physics II

– Macintosh and Windows versions available through PASCO Scientific, Admail 394, Cambridge CB1 1YY.

Suggested Projects

1. Compare the trajectories of a particle projected at an angle θ to the horizontal with speed u in vacuo and in air.

How will head/tail winds affect the trajectory in air? What difference will altitude make?

2. Compare the motion of a passenger riding in each of the following (a) an express lift (b) a car on a big dipper (c) a car on a loop-the-loop ride and (d) the Pirate Ship swing boat. On each of these rides, where in the path of the ride does the passenger experience the greatest "thrills" or physical effects? Can you predict where such points will be for other rides?

3. A satellite is in a circular orbit around the earth when it is given a small radial impulse outwards away from the centre of the earth through a brief firing of one of its control jets. What happens to the orbit of the satellite? What would happen to the orbit if the radial impulse were not small?

4. According to Einstein's "principle of equivalence" a uniform gravitational attraction is equivalent to a uniform acceleration. Find the inertial force which must be introduced if the equation of motion is to be written down relative to a freely falling frame of reference. Investigate whether such a freely falling frame of reference should be used as an inertial frame, rather than the frame $S_{\odot}$, when discussing terrestrial motions of small duration.

5. A three body system is moving under the action of their mutual gravitational attractions alone. Two of the bodies are of equal mass M and the mass m of the third body is much smaller than M. Discuss the motion of the light particle.

Solutions

Answers to Exercises at the end of Sections.

Exercises on 1.5

2 $\sqrt{l/g}$, $\sqrt{l'/g}$.

Exercises on 1.6

2 $y = 3x$, 1.0%.

Exercises on 1.7

1 $[MLT^{-2}]$, kms^{-2} **or N.**

2 $(x_2T_1 - x_1T_2)/(x_1 - x_2)$.

Exercises on 1.8

1 Stokes' drag is an appropriate model.

Exercises on 2.1

1 Acceleration of a sprinter is greater.

3 25s, 2.1s, 446m.

4 (a) 70 km per hour (b) (i) $2a\omega/\pi$ (ii) $2a\omega/\pi$.

5 294cms^{-2}, 0.12s.

Exercises on 2.2

1 5s (with $\delta t = 1$s).

2 1.4s.

3 0.7s, 5.3m.

Exercises on 2.3

1 $x_1 > x_2$, $x_1 < x_2$.

2 $x = \frac{F_0}{m\lambda^2}e^{-\lambda t} + v_0 t + \frac{F_0 t}{m\lambda} + x_0 - \frac{F_0}{m\lambda^2}$.

Exercises on 2.4

2 (i) $-\frac{1}{2}\lambda x^2 + \text{const}$ (ii) $\lambda/x + \text{const}$.

3 direction of motion must change.

Exercises on 2.5

2 oscillates if $\frac{1}{2}mv_0^2 < 2\lambda/27a$, escapes to $+\infty$ if $\frac{1}{2}mv_0^2 > 2\lambda/27a$ and approaches $x = 3a$ if $\frac{1}{2}mv_0^2 = 2\lambda/27a$.

3 (i) $u^2 < k/ma$ (ii) $k/ma < u^2 < 2k/ma$ (iii) $u^2 > 2k/ma$.

4 (a)$F = 2ax - 3bx^2$.

5 $V(x) = \frac{1}{2}k(x^2 + a^4/x^2)$, particle can remain at rest at either point of stable equilibrium, otherwise motion is oscillatory.

Exercises on 2.6

1 $(2l + g\tau^2)/2\tau$.

2 1.3m, 3.5ms^{-1}.

4 $d + g\tau^2$.

5 both speeds are $\frac{1}{2}g\tau$.

Exercises on 2.7

2 $1.4 \times 10^7\text{cms}^{-1}$, no.

3 speed decreases.

Exercises on 3.1

1 $AO = 3L$.

3 (iii) $0 < E < 1$ (v) ks^{-2} and $m^{-1}ks^{-2}$.

Exercises on 3.2

1 $\frac{1}{2\pi}\sqrt{\frac{k}{2ma^3}}$.

2 $A = 0.13\text{m}$, $\delta = 2.5\text{rad}$.

3 mg/x_0.

5 $\frac{1}{2\pi}\sqrt{\frac{\lambda}{ml_0}}$, $x = \mu l \sin(\omega t + \pi/2)$.

Exercises on 3.3

3 the same is true.

4 $k = -2(x_0 + v_0 T)/x_0 T$.

Exercises on 3.4

1 4s, 2.

3 $x = (F/2\omega)t \sin \omega t + A \sin(\omega t + \delta)$.

Exercises on 4.1

2 $(v, -v \cot \theta)$.

3 $(u + \sqrt{2v^2 - u^2}) : (-u + \sqrt{2v^2 - u^2})$.

4 563°E, 22 knots.

5 $\sqrt{f^2 + g^2}$.

Exercises on 4.4

3 kms^{-1}.

5 $m(0, 2b, 6ct^2), m(4bct^3, -6act^2, 2abt), m(bct^4, -2act^3, abt^2)$.

6 $(\lambda/2l_0)(r-l)^2 + \text{const}$.

Exercises on 5.1

1 no.

Exercises on 5.3

1 1.62ms^{-1}, $2.38 \times 10^3\text{ms}^{-1}$.

2 3 times the radius of the earth.

5 $\sqrt{\frac{R+H}{2GM}}\left[\sqrt{RH} + (R+H)\cos^{-1}\sqrt{\frac{R}{R+H}}\right]$.

Exercises on 6.1

1 $r = 5.9, \theta = 2.03\text{rad}$.

2 $(-2.1, -4.2)$.

Exercises on 6.2

1 $\omega d(e^{\omega t} - e^{-\omega t})/2,\ d(e^{\omega t} + e^{-\omega t})/2$.

2 $7.7 \times 10^3\text{ms}^{-1}$.

4 4.9ms^{-1}.

Exercises on 6.3

1 $T = ml\omega^2$.

2 $\sqrt{\lambda/ml_0}$.

Exercises on 6.4

1 0.003%.

2 $1.5 \times 10^{-5}\text{rad}$.

Exercises on 7.1

1 $L = m\omega\mathbf{b} \times \mathbf{a},\ E = \frac{1}{2}m\omega^2(|\mathbf{a}|^2 + |\mathbf{b}|^2)$.

Exercises on 7.2

6 a distance $\frac{Gm'}{v^2}\left(1 + \frac{4\delta v}{v}\right)$ from the sun, 180° beyond the boost.

Exercises on 7.3

1 $1.903 \times 10^{22}\text{kg}$.

2 26 years old, the year 2062.

Exercises on 7.4

1 $4.6 \times 10^6\text{m}$.

Index

Printed in the United Kingdom
by Lightning Source UK Ltd.
100134UKS00001B/329-346